The Hauptvermutung Book

## *K*-Monographs in Mathematics

VOLUME 1

This book series is devoted to developments in the mathematical sciences which have links to *K*-theory. Like the journal *K-theory*, it is open to all mathematical disciplines.

K-*Monographs in Mathematics* provides material for advanced undergraduate and graduate programmes, seminars and workshops, as well as for research activities and libraries.

The series' wide scope includes such topics as quantum theory, Kac-Moody theory, operator algebras, noncommutative algebraic and differential geometry, cyclic and related (co)homology theories, algebraic homotopy theory and homotopical algebra, controlled topology, Novikov theory, transformation groups, surgery theory, Hermitian and quadratic forms, arithmetic algebraic geometry, and higher number theory.

Researchers whose work fits this framework are encouraged to submit book proposals to the Series Editor or the Publisher.



*The titles published in this series are listed at the end of this volume.*

# The Hauptvermutung Book

## *A Collection of Papers of the Topology of Manifolds*

by

A.A. Ranicki (Editor)

*Department of Mathematics and Statistics,*
*The University of Edinburgh,*
*Edinburgh, Scotland*

A.J. Casson

*Mathematics Department,*
*University of California,*
*Berkeley, California, U.S.A.*

D.P. Sullivan

*Mathematics Department,*
*City University of New York,*
*New York, U.S.A.*

M.A. Armstrong

*Mathematics Department,*
*University of Durham,*
*Durham, U.K.*

C.P. Rourke

*Mathematics Institute,*
*University of Warwick,*
*Coventry, U.K.*

and

G.E. Cooke†

SPRINGER-SCIENCE+BUSINESS MEDIA, B.V.

A C.I.P. Catalogue record for this book is available from the Library of Congress.

ISBN 978-90-481-4735-9 ISBN 978-94-017-3343-4 (eBook)
DOI 10.1007/978-94-017-3343-4

*Printed on acid-free paper*

## Contents

**II. The Hauptvermutung according to Casson and Sullivan (by C.P.Rourke, 1968)**



**III. The Hauptvermutung according to Casson and Sullivan (by G.E.Cooke, 1968)**

# Preface

The Hauptvermutung is the conjecture that any two triangulations of a polyhedron are combinatorially equivalent. The conjecture was formulated at the turn of the century, and until its resolution was a central problem of topology. Initially, it was verified for low-dimensional polyhedra, and it might have been expected that further development of high-dimensional topology would lead to a verification in all dimensions. However, in 1961 Milnor constructed high-dimensional polyhedra with combinatorially inequivalent triangulations, disproving the Hauptvermutung in general. These polyhedra were not manifolds, leaving open the Hauptvermutung for manifolds. The development of surgery theory led to the disproof of the high-dimensional manifold Hauptvermutung in the late 1960's. Unfortunately, the published record of the manifold Hauptvermutung has been incomplete, as was forcefully pointed out by Novikov in his lecture at the Browder 60th birthday conference held at Princeton in March 1994.

This volume brings together the original 1967 papers of Casson and Sullivan, and the 1968/1972 'Princeton notes on the Hauptvermutung' of Armstrong, Rourke and Cooke, making this work physically accessible. These papers include several other results which have become part of the folklore but of which proofs have never been published. My own contribution is intended to serve as an introduction to the Hauptvermutung, and also to give an account of some more recent developments in the area.

In preparing the original papers for publication, only minimal changes of punctuation etc. have been made with the exception of the references, which have been completed and updated wherever possible.

I should like to thank all who in various ways helped in the preparation of this volume: Tony Armstrong, Tony Bak, Andrew Casson, Cathy Hassell, Bruce Hughes, Sergei Novikov, Jonathan Rosenberg, Colin Rourke, Ron Stern, Dennis Sullivan, Simon Willerton and Victor Wu.

A.A.R., Edinburgh, January 1996

# On the Hauptvermutung

by A. A. Ranicki

## §1. Introduction

An abstract simplicial complex $K$ determines a topological space, the polyhedron $|K|$. A **triangulation** $(K, f)$ of a topological space $X$ is a simplicial complex $K$ together with a homeomorphism $f : |K| \longrightarrow X$. A topological space $X$ is **triangulable** if it admits a triangulation $(K, f)$.

The topology of a triangulable space $X$ is determined by the combinatorial topology of the simplicial complex $K$ in any triangulation $(K, f)$ of $X$.

**Hauptvermutung** is short for **die Hauptvermutung der kombinatorischen Topologie**, which is German for the main conjecture of combinatorial topology. The conjecture states that the combinatorial topology of a simplicial complex $K$ is determined by the topology of the polyhedron $|K|$. More technically, the conjecture is that triangulations of homeomorphic spaces are combinatorially equivalent, i.e. become isomorphic after subdivision. A triangulable space would then have a canonical class of triangulations. The problem was formulated by Steinitz [44] and Tietze [48] in 1908, and there are statements in Kneser [20] and Alexandroff and Hopf [2, p.152].

The modern version of combinatorial topology is codified in the $PL$ (piecewise linear) category, for which Rourke and Sanderson [35] is the standard reference.

**Simplicial Approximation Theorem.** *Every continuous map $f : |K| \longrightarrow |L|$ between polyhedra is homotopic to the topological realization $|f'| : |K| = |K'| \longrightarrow |L|$ of a simplicial map $f' : K' \longrightarrow L$, where $K'$ is a simplicial subdivision of $K$.*

Thus every continuous map of polyhedra is homotopic to a $PL$ map. It does not follow that a homeomorphism of polyhedra is homotopic to a $PL$ homeomorphism.

**Hauptvermutung.** *Every homeomorphism $f : |K| \longrightarrow |L|$ between polyhedra is homotopic to the topological realization of a simplicial isomorphism $f' : K' \longrightarrow L'$, where $K', L'$ are simplicial subdivisions of $K, L$, i.e. every homeomorphism of polyhedra is homotopic to a $PL$ homeomorphism.*

This will also be called the **Polyhedral Hauptvermutung**, to distinguish it from the Manifold Hauptvermutung stated below.

The Simplicial Approximation Theorem shows that the homotopy theory of polyhedra is the same as the $PL$ homotopy theory of simplicial complexes. Ever since Seifert and Threlfall [39] standard treatments of algebraic topology have used this correspondence to show that combinatorial homotopy invariants of simplicial complexes (e.g. simplicial homology, the simplicial fundamental group) are in fact homotopy invariants of polyhedra. The Hauptvermutung is not mentioned, allowing the reader to gain the false impression that the topology of polyhedra is the same as the $PL$ topology of simplicial complexes. In fact, the Hauptvermutung has been known for some time to be false, although this knowledge has not yet filtered down to textbook level.

A simplicial complex $K$ is finite if and only if the polyhedron $|K|$ is compact. The Hauptvermutung is only considered here for compact polyhedra. However, the resolution of the conjecture in this case requires an understanding of the difference between the $PL$ and continuous topology of open $PL$ manifolds, which is quantified by the Whitehead group.

The Polyhedral Hauptvermutung is true in low dimensions: it was verified for 2-dimensional manifolds by the classification of surfaces, for all polyhedra of dimension $\leq 2$ by Papakyriakopoulos [32], and by Moïse [28] for 3-dimensional manifolds.

Milnor [25] obtained the first counterexamples to the Polyhedral Hauptvermutung in 1961, using Reidemeister torsion and some results on non-compact manifolds of Mazur and Stallings to construct a homeomorphism of compact polyhedra which is not homotopic to a $PL$ homeomorphism. Stallings [43] generalized the construction, showing that any non-trivial $h$-cobordism determines a counterexample to the Polyhedral Hauptvermutung. These counterexamples arise as homeomorphisms of one-point compactifications of open $PL$ manifolds, and so are non-manifold in nature.

An $m$-**dimensional combinatorial (or** $PL$**) manifold** is a simplicial complex $K$ such that $\text{link}_K(\sigma)$ is a $PL$ $(m - |\sigma| - 1)$-sphere for each simplex $\sigma \in K$.

**Manifold Hauptvermutung.** *Every homeomorphism $f : |K| \longrightarrow |L|$ of the polyhedra of compact $m$-dimensional $PL$ manifolds is homotopic to a $PL$ homeomorphism.*

Following Milnor's disproof of the Polyhedral Hauptvermutung there was much activity in the 1960's aimed at the Manifold Hauptvermutung – first proving it in special cases, and then disproving it in general.

The Manifold Hauptvermutung is the rel $\partial$ version of the following conjecture:

**Combinatorial Triangulation Conjecture.** *Every compact m-dimensional topological manifold M can be triangulated by a PL manifold.*

The Manifold Hauptvermutung and Combinatorial Triangulation Conjecture hold in the low dimensions $m \leq 3$.

The 1963 surgery classification by Kervaire and Milnor of the latter's exotic differentiable spheres led to smoothing theory, which gave a detailed understanding of the relationship between differentiable and $PL$ manifold structures. The subsequent Browder-Novikov-Sullivan-Wall surgery theory of high-dimensional manifolds was initially applied to differentiable and $PL$ manifolds. The theory deals with the homotopy analogues of the Manifold Hauptvermutung and the Combinatorial Triangulation Conjecture, providing the necessary and sufficient algebraic topology to decide whether a homotopy equivalence of $m$-dimensional $PL$ manifolds $f : K \longrightarrow L$ is homotopic to a $PL$ homeomorphism, and whether an $m$-dimensional Poincaré duality space is homotopy equivalent to an $m$-dimensional $PL$ manifold, at least for $m \geq 5$. The 1965 proof by Novikov [31] of the topological invariance of the rational Pontrjagin classes ultimately made it possible to extend the theory to topological manifolds and homeomorphisms, and to resolve the Manifold Hauptvermutung and the Combinatorial Triangulation Conjecture using algebraic $K$- and $L$-theory.

In 1969 the surgery classification of $PL$ structures on high-dimensional tori allowed Kirby and Siebenmann to show that the Manifold Hauptvermutung and the Combinatorial Triangulation Conjecture are false in general, and to extend high-dimensional surgery theory to topological manifolds. The book of Kirby and Siebenmann [19] is the definitive account of their work. Some of the late 1960's original work on the Manifold Hauptvermutung was announced at the time, e.g. Sullivan [46], [47], Lashof and Rothenberg [21], Kirby and Siebenmann [18], Siebenmann [42]. However, not all the results obtained have been published. The 1967 papers of Casson [5] and Sullivan [45] are published in this volume, along with the 1968/1972 notes of Armstrong et. al. [3].

Kirby and Siebenmann used the Rochlin invariant to formulate an invariant $\kappa(M) \in H^4(M;\mathbb{Z}_2)$ for any closed topological manifold $M$, such that, for $\dim(M) \geq 5$, $\kappa(M) = 0$ if and only if $M$ admits a combinatorial triangulation. A homeomorphism $f : |K| \longrightarrow |L|$ of the polyhedra of closed $PL$ manifolds gives rise to an invariant $\kappa(f) \in H^3(L;\mathbb{Z}_2)$ (the rel $\partial$ combinatorial triangulation obstruction of the mapping cylinder) such that for $\dim(L) \geq 5$ $\kappa(f) = 0$ if and only if $f$ is homotopic to a $PL$ homeomorphism. These obstructions are realized. For $m \geq 5$ and any element $\kappa \in H^3(T^m;\mathbb{Z}_2)$ there exists a combinatorial triangulation $(\tau^m, f)$ of $T^m$ with $\kappa(f) = \kappa$, so that for $\kappa \neq 0$ the homeomorphism $f : \tau^m \longrightarrow T^m$ is not homotopic to a $PL$ homeomorphism. For $m \geq 3$, $k \geq 2$ and any $\kappa \in H^4(T^m;\mathbb{Z}_2)$ there exists a closed $(m+k)$-dimensional topological manifold $M$ with a homotopy equivalence $h: M \longrightarrow T^m \times S^k$ such that $\kappa(M) = h^*\kappa$, so that for $\kappa \neq 0$ $M$ does not admit a combinatorial triangulation. Such counterexamples to the Manifold Hauptvermutung and the Combinatorial Triangulation Conjecture in dimensions $\geq 5$ can be traced to the 3-dimensional Poincaré homology sphere $\Sigma$. See §§3-5 for a more detailed account of the Kirby-Siebenmann obstruction.

## §2. The Polyhedral Hauptvermutung

**Theorem.** (Milnor [25]) *The Polyhedral Hauptvermutung is false: there exists a homeomorphism $f: |K| \longrightarrow |L|$ of the polyhedra of finite simplicial complexes $K, L$ such that $f$ is not homotopic to a $PL$ homeomorphism.*

The failure of the Polyhedral Hauptvermutung is detected by Whitehead torsion. The construction of the Polyhedral Hauptvermutung counterexamples of Milnor [25] and Stallings [43] will now be recounted, first directly and then using the end obstruction theory of Siebenmann [40]. See Cohen [7] for a textbook account.

Given a topological space $X$ let

$$X^\infty = X \cup \{\infty\}$$

be the one-point compactification. If $X$ is compact then $X^\infty$ is just the union of $X$ and $\{\infty\}$ as topological spaces.

Let $(W, \partial W)$ be a compact $n$-dimensional topological manifold with non-empty boundary $\partial W$, so that the interior

$$\dot{W} = W \backslash \partial W$$

is an open $n$-dimensional manifold. Since $\partial W$ is collared in $W$ (i.e. the inclusion $\partial W = \partial W \times \{0\} \longrightarrow W$ extends to an embedding $\partial W \times [0,1] \longrightarrow W$) the effect of collapsing the boundary to a point is a compact space

$$K = W/\partial W$$

with a homeomorphism

$$K \cong \dot{W}^\infty$$

sending $[\partial W] \in K$ to $\infty$. Now suppose that $(W, \partial W)$ is a $PL$ manifold with boundary, so that $\dot{W}$ is an open $n$-dimensional $PL$ manifold, and $K$ is a compact polyhedron such that there is defined a $PL$ homeomorphism

$$\mathrm{link}_K(\infty) \cong \partial W .$$

The compact polyhedron $K$ is a closed $n$-dimensional $PL$ manifold if and only if $\partial W$ is a $PL$ $(n-1)$-sphere. If $\partial W$ is not a $PL$ $(n-1)$-sphere then $K$ is a $PL$

stratified set with two strata, $\dot{W}$ and $\{\infty\}$.

Suppose given compact $n$-dimensional $PL$ manifolds with boundary $(W_1, \partial W_1)$, $(W_2, \partial W_2)$ such that

$$W_2 = W_1 \cup_{\partial W_1} V$$

for an $h$-cobordism $(V; \partial W_1, \partial W_2)$.

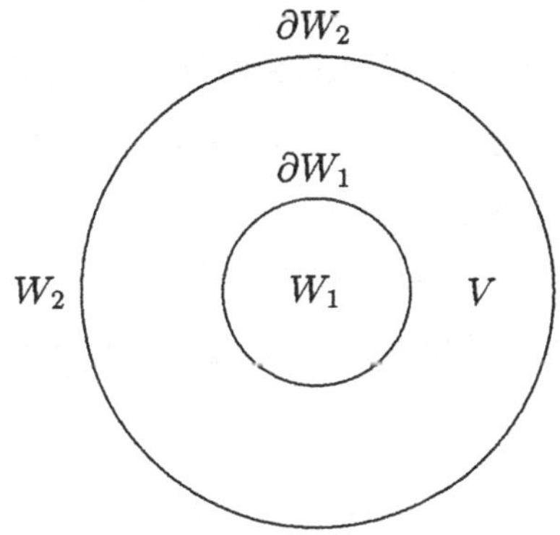


There is defined a $PL$ homeomorphism

$$(V \backslash \partial W_2, \partial W_1) \cong \partial W_1 \times ([0,1), \{0\})$$

of non-compact $n$-dimensional $PL$ manifolds with boundary, which is the identity on $\partial W_1$. The corresponding $PL$ homeomorphism of open $n$-dimensional $PL$ manifolds $\dot{W}_1 \longrightarrow \dot{W}_2$ compactifies to a homeomorphism of compact polyhedra

$$f : K_1 = W_1/\partial W_1 = \dot{W}_1^\infty \longrightarrow K_2 = W_2/\partial W_2 = \dot{W}_2^\infty .$$

The homeomorphism $f$ is homotopic to a $PL$ homeomorphism if and only if there exists a $PL$ homeomorphism

$$(V; \partial W_1, \partial W_2) \cong \partial W_1 \times ([0,1]; \{0\}, \{1\})$$

which is the identity on $\partial W_1$.

If $M$ is a closed $m$-dimensional $PL$ manifold then for any $i \geq 1$

$$(W, \partial W) = M \times (D^i, S^{i-1})$$

is a compact $(m+i)$-dimensional $PL$ manifold with boundary such that

$$\begin{aligned} \dot{W} &= M \times \mathbb{R}^i , \\ W/\partial W &= \dot{W}^\infty = M \times D^i / M \times S^{i-1} = \Sigma^i M^\infty . \end{aligned}$$

Milnor [25] applied this construction to obtain the first counterexamples to the Hauptvermutung, using the Reidemeister-Franz-deRham-Whitehead classification of the lens spaces introduced by Tietze [48]. The lens spaces $L(7,1)$, $L(7,2)$ are closed 3-dimensional $PL$ manifolds which are homotopy equivalent but not simple homotopy equivalent, and hence neither $PL$ homeomorphic nor homeomorphic (by the topological invariance of Whitehead torsion). For $i \geq 3$ the compact $(i+3)$-dimensional $PL$ manifolds with boundary

$$\begin{aligned}(W_1, \partial W_1) &= L(7,1) \times (D^i, S^{i-1}) , \\ (W_2, \partial W_2) &= L(7,2) \times (D^i, S^{i-1})\end{aligned}$$

are such that $W_2 = W_1 \cup_{\partial W_1} V$ for an $h$-cobordism $(V; \partial W_1, \partial W_2)$ with torsion

$$\tau(\partial W_1 \longrightarrow V) \neq 0 \in Wh(\mathbb{Z}_7) = \mathbb{Z} \oplus \mathbb{Z} .$$

(See Milnor [26] for $Wh(\mathbb{Z}_7)$.) The corresponding $PL$ homeomorphism of open $(i+3)$-dimensional $PL$ manifolds

$$\dot{W}_1 = L(7,1) \times \mathbb{R}^i \longrightarrow \dot{W}_2 = L(7,2) \times \mathbb{R}^i$$

compactifies to a homeomorphism of compact polyhedra

$$f : K_1 = \Sigma^i L(7,1)^\infty \longrightarrow K_2 = \Sigma^i L(7,2)^\infty$$

which is not homotopic to a $PL$ homeomorphism. In fact, $f$ is homotopic to the $i$-fold suspension of a homotopy equivalence $h : L(7,1) \longrightarrow L(7,2)$ with Whitehead torsion

$$\begin{aligned}\tau(h) &= \tau(\partial W_1 \longrightarrow V) + \tau(\partial W_1 \longrightarrow V)^* \\ &= 2\tau(\partial W_1 \longrightarrow V) \neq 0 \in Wh(\mathbb{Z}_7) .\end{aligned}$$

The homotopy equivalence $h$ is not homotopic to a homeomorphism (by the topological invariance of Whitehead torsion) let alone a $PL$ homeomorphism.

Let $(N, \partial N)$ be a non-compact $n$-dimensional $PL$ manifold with a compact boundary $\partial N$ and a tame end $\epsilon$.

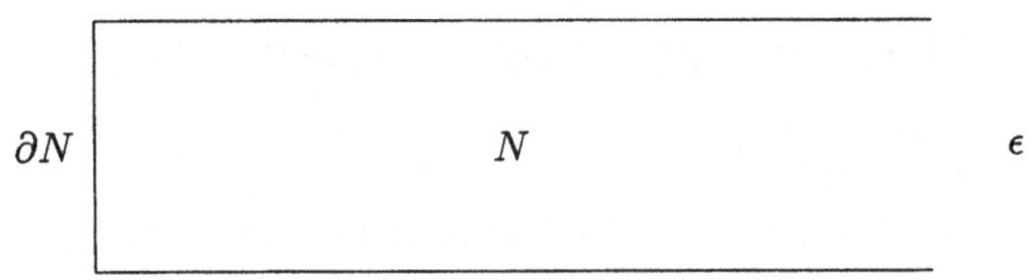


A **closure** of the tame end $\epsilon$ is a compact $n$-dimensional $PL$ cobordism $(W; \partial N, \partial_+ W)$

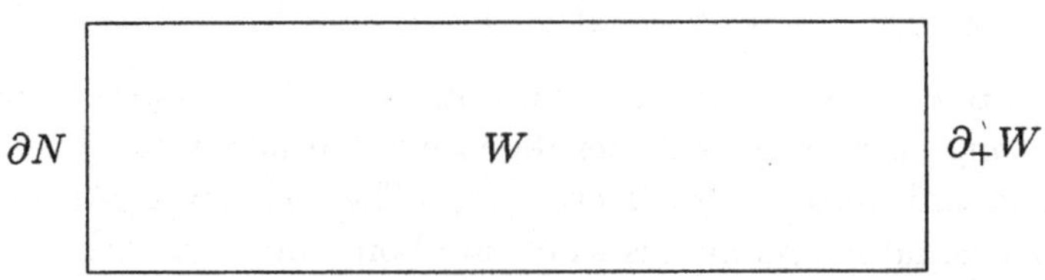

with a $PL$ homeomorphism

$$N \cong W\backslash\partial_+ W$$

which is the identity on $\partial N$, in which case $\pi_1(\partial_+ W) = \pi_1(\epsilon)$ and there is defined a homeomorphism

$$W/\partial_+ W \cong N^\infty .$$

The **end obstruction** $[\epsilon] \in \widetilde{K}_0(\mathbb{Z}[\pi_1(\epsilon)])$ of Siebenmann [40] is such that $[\epsilon] = 0$ if (and for $n \geq 6$ only if) $\epsilon$ admits a closure. The end obstruction has image the Wall finiteness obstruction $[N] \in \widetilde{K}_0(\mathbb{Z}[\pi_1(N)])$, which is such that $[N] = 0$ if and only if $N$ is homotopy equivalent to a compact polyhedron. See Hughes and Ranicki [16] for a recent account of tame ends and the end obstruction.

The closures of high-dimensional tame ends $\epsilon$ are classified by the Whitehead group $Wh(\pi_1(\epsilon))$. This is a consequence of:

**$s$-cobordism Theorem.** (Barden, Mazur, Stallings)
*An $n$-dimensional $PL$ $h$-cobordism $(V;U,U')$ with torsion*

$$\tau(U \longrightarrow V) = \tau \in Wh(\pi_1(U))$$

*is such that $\tau = 0$ if (and for $n \geq 6$ only if) there exists a $PL$ homeomorphism*

$$(V;U,U') \cong U \times ([0,1];\{0\},\{1\})$$

*which is the identity on $U$.*

Let $(W_1;\partial N,\partial_+ W_1)$, $(W_2;\partial N,\partial_+ W_2)$ be two closures of an $n$-dimensional tame end $\epsilon$ (as above), so that there are defined $PL$ homeomorphisms

$$N \cong W_1\backslash\partial_+ W_1 \cong W_2\backslash\partial_+ W_2$$

and a homeomorphism of compact polyhedra

$$f : K_1 = W_1/\partial_+ W_1 \longrightarrow K_2 = W_2/\partial_+ W_2 .$$

The points

$$\infty_1 = [\partial_+ W_1] \in K_1 \ , \ \infty_2 = [\partial_+ W_2] \in K_2$$

are such that

$$\mathrm{link}_{K_1}(\infty_1) = \partial_+ W_1 \ , \ \mathrm{link}_{K_2}(\infty_2) = \partial_+ W_2 .$$

If neither $\partial_+ W_1$ nor $\partial_+ W_2$ is a $PL$ $(n-1)$-sphere then these are the only non-manifold points of $K_1, K_2$ – any $PL$ homeomorphism $F : K_1 \longrightarrow K_2$ would have to be such that $F(\infty_1) = \infty_2$ and restrict to a $PL$ homeomorphism

$$F : \mathrm{link}_{K_1}(\infty_1) = \partial_+ W_1 \longrightarrow \mathrm{link}_{K_2}(\infty_2) = \partial_+ W_2 .$$

If $\partial_+ W_1$ is not $PL$ homeomorphic to $\partial_+ W_2$ there will not exist such an $F$ which provides a counterexample to the Hauptvermutung. In any case, for $n \geq 6$ there exists an $n$-dimensional $PL$ $h$-cobordism $(V;\partial_+W_1,\partial_+W_2)$ such that up to $PL$ homeomorphism

$$(W_2;\partial N,\partial_+W_2) \ = \ (W_1;\partial N,\partial_+W_1)\cup(V;\partial_+W_1,\partial_+W_2)$$

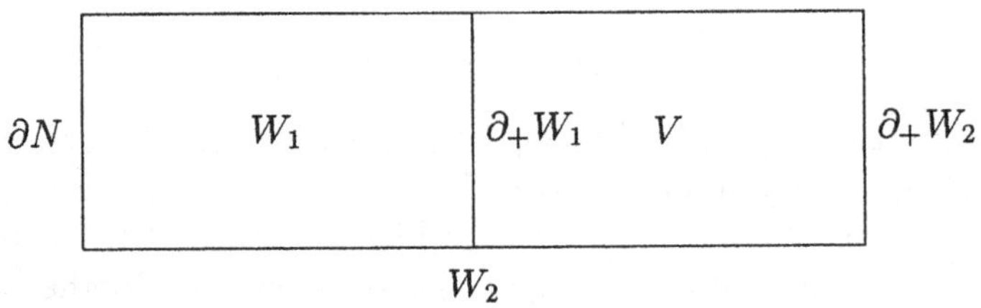


and the following conditions are equivalent :

(i) the Whitehead torsion

$$\tau \ = \ \tau(\partial_+W_1 \longrightarrow V) \in Wh(\pi_1(V)) \ = \ Wh(\pi_1(\epsilon))$$

is such that $\tau = 0$,

(ii) there exists a $PL$ homeomorphism

$$(W_1;\partial N,\partial_+W_1) \ \cong \ (W_2;\partial N,\partial_+W_2)$$

which is the identity on $\partial N$,

(iii) there exists a $PL$ homeomorphism

$$(V;\partial_+W_1,\partial_+W_2) \ \cong \ \partial_+W_1\times([0,1];\{0\},\{1\})$$

which is the identity on $\partial_+W_1$,

(iv) the homeomorphism $f : K_1 \longrightarrow K_2$ is homotopic to a $PL$ homeomorphism.

Returning to the construction of Milnor [25], define for any $i \geq 1$ the open $(i+3)$-dimensional $PL$ manifold

$$N \ = \ L(7,1)\times\mathbb{R}^i$$

with a tame end $\epsilon$, which can be closed in the obvious way by

$$(W_1,\partial W_1) \ = \ L(7,1)\times(D^i,S^{i-1}) \ .$$

For $i \geq 3$ use the above $h$-cobordism $(V;\partial W_1,\partial W_2)$ with $\tau(\partial W_1 \longrightarrow V) \neq 0$ to close $\epsilon$ in a non-obvious way, with $W_2 = W_1 \cup_{\partial W_1} V$ such that

$$(W_2,\partial W_2) \ = \ L(7,2)\times(D^i,S^{i-1}) \ ,$$

and as before there is a homeomorphism of compact polyhedra

$$f \ : \ K_1 \ = \ W_1/\partial W_1 \ = \ \Sigma^i L(7,1)^\infty \longrightarrow K_2 \ = \ W_2/\partial W_2 \ = \ \Sigma^i L(7,2)^\infty$$

which is not homotopic to a $PL$ homeomorphism.

A closed $m$-dimensional $PL$ manifold $M$ determines a non-compact $(m+1)$-dimensional $PL$ manifold with compact boundary

$$(N,\partial N) \ = \ (M\times[0,1),M\times\{0\})$$

with a tame end $\epsilon$ which can be closed, so that

$$[\epsilon] = 0 \in \widetilde{K}_0(\mathbb{Z}[\pi_1(\epsilon)]) = \widetilde{K}_0(\mathbb{Z}[\pi_1(M)]) .$$

Assume $m \geq 5$. For any $(m+1)$-dimensional $PL$ $h$-cobordism $(W; M, M')$ the inclusion $M \subset W \backslash M'$ extends to a $PL$ homeomorphism of open $PL$ manifolds

$$U = M \times [0,1) \longrightarrow W \backslash M'$$

which compactifies to a homeomorphism of compact polyhedra

$$f : K = U^\infty = M \times [0,1]/M \times \{1\} \longrightarrow L = W/M'$$

such that $f$ is homotopic to a $PL$ homeomorphism if and only if $\tau = 0 \in Wh(\pi_1(M))$. Thus any $h$-cobordism with $\tau \neq 0$ determines a counterexample to the Polyhedral Hauptvermutung, as was first observed by Stallings [43].

## §3. The Rochlin Invariant

The **intersection form** of a compact oriented $4k$-dimensional manifold with boundary $(M, \partial M)$ is the symmetric form

$$\phi : H^{2k}(M, \partial M) \times H^{2k}(M, \partial M) \longrightarrow \mathbb{Z} \; ; \; (x, y) \longmapsto \langle x \cup y, [M] \rangle$$

where $[M] \in H_{4k}(M, \partial M)$ is the fundamental class. The **signature** of $(M, \partial M)$ is

$$\sigma(M) = \text{signature}(H^{2k}(M, \partial M), \phi) \in \mathbb{Z} .$$

**Proposition.** *Let $M$ be a closed oriented 4-dimensional topological manifold. For any integral lift $w_2 \in H^2(M)$ of the second Stiefel-Whitney class $w_2(M) \in H^2(M; \mathbb{Z}_2)$*

$$\phi(x, x) \equiv \phi(x, w_2) \pmod 2 \quad (x \in H^2(M)) ,$$

*and*

$$\sigma(M) \equiv \phi(w_2, w_2) \pmod 8 .$$

**Proof.** See Milnor and Husemoller [27, II.5].

A closed oriented 4-dimensional manifold $M$ is **spin** (i.e. admits a spin structure) if $w_2(M) = 0 \in H^2(M; \mathbb{Z}_2)$, in which case $\sigma(M) \equiv 0 (\bmod 8)$.

**Theorem.** (Rochlin) *The signature of a closed oriented 4-dimensional PL spin manifold $M$ is such that*

$$\sigma(M) \equiv 0 \pmod{16} .$$

**Proof.** See Guillou and Marin [13], Kirby [17, XI].

**Definition.** (i) The **Rochlin invariant** of a closed oriented 4-dimensional topological spin manifold $M$ is

$$\alpha(M) = \sigma(M) \in 8\mathbb{Z}/16\mathbb{Z} = \mathbb{Z}_2 .$$

(ii) The **Rochlin invariant** of an oriented 3-dimensional $PL$ homology sphere $\Sigma$ is defined by

$$\alpha(\Sigma) = \sigma(W) \in 8\mathbb{Z}/16\mathbb{Z} = \mathbb{Z}_2 ,$$

for any 4-dimensional $PL$ spin manifold $(W, \partial W)$ with boundary $\partial W = \Sigma$.

**Proposition.** (i) *Let $(M, \partial M)$ be a connected 4-dimensional topological spin manifold with homology 3-sphere boundary $\partial M = \Sigma$. The Rochlin invariant of $\Sigma$ is expressed in terms of the signature of $M$ and the Kirby-Siebenmann invariant of the stable normal bundle $\nu_M : M \longrightarrow BTOP$ by*

$$\alpha(\Sigma) = \sigma(M)/8 - \kappa(M) \in H^4(M, \partial M; \mathbb{Z}_2) = \mathbb{Z}_2 .$$

(ii) *Let $M$ be a connected closed oriented 4-dimensional topological spin manifold. The Rochlin invariant of $M$ is the Kirby-Siebenmann invariant of $M$*

$$\alpha(M) = \kappa(M) \in H^4(M; \mathbb{Z}_2) = \mathbb{Z}_2 .$$

**Proof.** (i) By Freedman and Quinn [10, 10.2B], for any 4-dimensional topological spin manifold with boundary $(M, \partial M)$, there exists a 4-dimensional $PL$ spin manifold with boundary $(N, \partial N)$ such that $\partial M = \partial N$, and for any such $M, N$

$$\tfrac{1}{8}(\sigma(M) - \sigma(N)) = \kappa(M) \in H^4(M, \partial M; \mathbb{Z}_2) = \mathbb{Z}_2$$

(ii) Take $\partial M = \emptyset$, $N = \emptyset$ in (i).

**Examples.** (i) The Poincaré homology 3-sphere

$$\Sigma = SO(3)/A_5$$

is the boundary of the parallelizable 4-dimensional $PL$ manifold $Q$ obtained by Milnor's $E_8$-plumbing, such that

$$\sigma(Q) = 8 \in \mathbb{Z} , \ \alpha(\Sigma) = 1 \in \mathbb{Z}_2 , \ \kappa(Q) = 0 \in \mathbb{Z}_2 .$$

(ii) Any 3-dimensional topological manifold $\Sigma$ with the homology of $S^3$ bounds a contractible topological 4-manifold (Freedman and Quinn [10, 9.3C]). If $(Q, \Sigma)$ is as in (i) and $W$ is a contractible topological 4-manifold with boundary $\partial W = \Sigma$ there is obtained the Freedman $E_8$-manifold $M = W \cup_\Sigma Q$, a closed oriented 4-dimensional topological spin manifold such that

$$\sigma(M) = 8 \in \mathbb{Z} , \ \alpha(M) = 1 \in \mathbb{Z}_2 , \ \kappa(M) = 1 \in \mathbb{Z}_2 .$$

## §4. The Manifold Hauptvermutung

**Theorem.** (Kirby-Siebenmann [18], [19])
*The Combinatorial Triangulation Conjecture and the Manifold Hauptvermutung are false in each dimension $m \geq 5$: there exist compact $m$-dimensional topological manifolds without a PL structure (= combinatorial triangulation), and there exist homeomorphisms $f : |K| \longrightarrow |L|$ of the polyhedra of compact $m$-dimensional PL manifolds $K, L$ which are not homotopic to a PL homeomorphism.*

The actual construction of counterexamples required the surgery classification of homotopy tori due to Wall, Hsiang and Shaneson, and Casson using the non-simply-connected surgery theory of Wall [49]. The failure of the Combinatorial Triangulation Conjecture is detected by the Kirby-Siebenmann invariant, which uses the Rochlin invariant to detect the difference between topological and $PL$ bundles. The failure of the Manifold Hauptvermutung is detected by the Casson-Sullivan invariant, which is the rel $\partial$ version of the Kirby-Siebenmann invariant. For $m \geq 5$ an $m$-dimensional topological manifold admits a combinatorial triangulation if and only if the stable normal topological bundle admits a $PL$ bundle refinement. A homeomorphism of $m$-dimensional $PL$ manifolds is homotopic to a $PL$ homeomorphism if and only if it preserves the stable normal $PL$ bundles.

A stable topological bundle $\eta$ over a compact polyhedron $X$ is classified by the homotopy class of a map

$$\eta : X \longrightarrow BTOP$$

to a classifying space

$$BTOP = \varinjlim_{k} BTOP(k) .$$

There is a similar result for $PL$ bundles. The classifying spaces $BTOP, BPL$ are related by a fibration sequence

$$TOP/PL \longrightarrow BPL \longrightarrow BTOP \longrightarrow B(TOP/PL) .$$

A stable topological bundle $\eta : X \longrightarrow BTOP$ lifts to a stable $PL$ bundle $\widetilde{\eta} : X \longrightarrow BPL$ if and only if the composite

$$X \xrightarrow{\eta} BTOP \longrightarrow B(TOP/PL)$$

is null-homotopic.

**Theorem.** (Kirby-Siebenmann [18], [19] for $m \geq 5$, Freedman-Quinn [10] for $m = 4$)
(i) *There is a homotopy equivalence*

$$B(TOP/PL) \simeq K(\mathbb{Z}_2, 4) .$$

*Given a stable topological bundle* $\eta : X \longrightarrow BTOP$ *let*

$$\kappa(\eta) \in [X, B(TOP/PL)] = H^4(X; \mathbb{Z}_2)$$

*be the homotopy class of the composite* $X \xrightarrow{\eta} BTOP \longrightarrow B(TOP/PL)$. *The topological bundle* $\eta$ *lifts to a stable PL bundle* $\widetilde{\eta} : X \longrightarrow BPL$ *if and only if* $\kappa(\eta) = 0$.
(ii) *There is a homotopy equivalence*

$$TOP/PL \simeq K(\mathbb{Z}_2, 3) .$$

*A topological trivialization* $t : \widetilde{\eta} \simeq \{*\} : X \longrightarrow BTOP$ *of a stable PL bundle* $\widetilde{\eta} : X \longrightarrow BPL$ *corresponds to a lift of* $\widetilde{\eta}$ *to a map* $(\widetilde{\eta}, t) : X \longrightarrow TOP/PL$. *It is possible to further refine* $t$ *to a PL trivialization if and only if the homotopy class*

$$\kappa(\widetilde{\eta}, t) \in [X, TOP/PL] = H^3(X; \mathbb{Z}_2)$$

*is such that* $\kappa(\widetilde{\eta}, t) = 0$.
(iii) *The* **Kirby-Siebenmann invariant** *of a compact* $m$*-dimensional topological manifold* $M$ *with a PL boundary* $\partial M$ *(which may be empty)*

$$\kappa(M) = \kappa(\nu_M : M \longrightarrow BTOP) \in H^4(M, \partial M; \mathbb{Z}_2)$$

*is such that* $\kappa(M) = 0 \in H^4(M, \partial M; \mathbb{Z}_2)$ *if and only if there exists a PL reduction* $\widetilde{\nu}_M : M \longrightarrow BPL$ *of* $\nu_M : M \longrightarrow BTOP$ *which extends* $\nu_{\partial M} : \partial M \longrightarrow BPL$. *The invariant is such that* $\kappa(M) = 0$ *if (and for* $m \geq 4$ *only if) the PL structure on* $\partial M$ *extends to a PL structure on* $M \times \mathbb{R}$. *For* $m \geq 5$ *such a PL structure on* $M \times \mathbb{R}$ *is determined by a PL structure on* $M$.
(iv) *Let* $f : |K| \longrightarrow |L|$ *be a homeomorphism of the polyhedra of closed* $m$*-dimensional PL manifolds. The mapping cylinder*

$$W = |K| \times I \cup_f |L|$$

*is an* $(m+1)$*-dimensional topological manifold with PL boundary* $\partial W = |K| \times \{0\} \cup |L|$. *The* **Casson-Sullivan invariant** *of* $f$ *is defined by*

$$\kappa(f) = \kappa(W) \in H^4(W, \partial W; \mathbb{Z}_2) = H^3(L; \mathbb{Z}_2) .$$

*For* $m \geq 4$ *the following conditions are equivalent:*

(a) $f$ *is homotopic to a PL homeomorphism,*
(b) $W$ *has a PL structure extending the PL structure on* $\partial W$,
(c) $\kappa(f) = 0 \in H^3(L; \mathbb{Z}_2)$.

(v) *The Combinatorial Triangulation Conjecture is false for* $m \geq 4$: *there exist closed* $m$*-dimensional topological manifolds* $M$ *such that* $\kappa \neq 0 \in H^4(M; \mathbb{Z}_2)$, *which thus do not admit a combinatorial triangulation.*

(vi) *The Manifold Hauptvermutung is false for* $m \geq 4$ *: for every closed* $m$*-dimensional PL manifold* $L$ *and every* $\kappa \in H^3(L;\mathbb{Z}_2)$ *there exists a closed* $m$*-dimensional PL manifold* $K$ *with a homeomorphism* $f : |K| \longrightarrow |L|$ *such that* $\kappa(f) = \kappa \in H^3(L;\mathbb{Z}_2)$.

The stable classifying spaces $BPL, BTOP, BG$ for bundles in the $PL$, topological and homotopy categories are related by a braid of fibrations

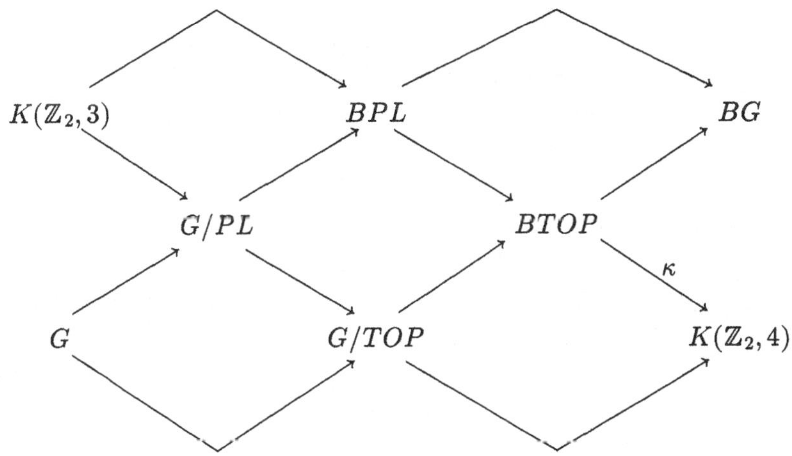


Sullivan determined the homotopy types of the surgery classifying spaces $G/PL$ and $G/TOP$. See Madsen and Milgram [22] for an account of this determination, and Rudyak [36] for an account of its application to the Manifold Hauptvermutung.

The homotopy groups of $G/TOP$ are the simply-connected surgery obstruction groups

$$\pi_m(G/TOP) = L_m(\mathbb{Z}) = \begin{cases} \mathbb{Z} & \text{if } m \equiv 0 (\bmod 4) \\ \mathbb{Z}_2 & \text{if } m \equiv 2 (\bmod 4) \\ 0 & \text{if } m \equiv 1,3 (\bmod 4)\ . \end{cases}$$

A map $(\eta, t) : S^m \longrightarrow G/TOP$ corresponds to a topological bundle $\eta : S^m \longrightarrow BTOP(k)$ ($k$ large) with a fibre homotopy trivialization $t : \eta \simeq \{*\} : S^m \longrightarrow BG(k)$. Making the degree 1 map $\rho : S^{m+k} \longrightarrow T(\eta)$ topologically transverse regular at $S^m \subset T(\eta)$ gives an $m$-dimensional normal map by the Browder-Novikov construction

$$(f,b) = \rho| : M^m = \rho^{-1}(S^m) \longrightarrow S^m$$

with $b : \nu_M \longrightarrow \eta$. The homotopy class of $(\eta, t)$ is the surgery obstruction of $(f,b)$

$$(\eta,t) = \sigma_*(f,b) \in \pi_m(G/TOP) = L_m(\mathbb{Z})\ ,$$

where

$$\sigma_*(f,b) \;=\; \begin{cases} \frac{1}{8}\sigma(M) \in L_{4k}(\mathbb{Z}) \;=\; \mathbb{Z} & \text{if } m = 4k \\ c(M) \in L_{4k+2}(\mathbb{Z}) \;=\; \mathbb{Z}_2 & \text{if } m = 4k+2 \end{cases}$$

with $c(M) \in \mathbb{Z}_2$ the Kervaire-Arf invariant of the framed $(4k+2)$-dimensional manifold $M$. Similarly for maps $(\widetilde{\eta},t) : S^m \longrightarrow G/PL$.

The low-dimensional homotopy groups of the bundle classifying spaces are given by

$$\pi_m(BPL) \;=\; \pi_m(BO) \;=\; \begin{cases} \mathbb{Z}_2 & \text{if } m = 1,2 \\ 0 & \text{if } m = 3,5,6,7 \\ \mathbb{Z} & \text{if } m = 4 \end{cases}$$

$$\pi_m(BTOP) \;=\; \begin{cases} \mathbb{Z}_2 & \text{if } m = 1,2 \\ 0 & \text{if } m = 3,5,6,7 \\ \mathbb{Z}\oplus\mathbb{Z}_2 & \text{if } m = 4\ . \end{cases}$$

The first Pontrjagin class $p_1(\eta) \in H^4(S^4) = \mathbb{Z}$ and the Kirby-Siebenmann invariant $\kappa(\eta) \in H^4(S^4;\mathbb{Z}_2) = \mathbb{Z}_2$ define isomorphisms

$$\pi_4(BPL) \xrightarrow{\cong} \mathbb{Z}\ ;\ (\widetilde{\eta} : S^4 \longrightarrow BPL) \longmapsto \tfrac{1}{2}p_1(\widetilde{\eta})\ ,$$

$$\pi_4(BTOP) \xrightarrow{\cong} \mathbb{Z}\oplus\mathbb{Z}_2\ ;\ (\eta : S^4 \longrightarrow BTOP) \longmapsto (p_1(\eta),\kappa(\eta))\ .$$

For any map $(\eta,t) : S^4 \longrightarrow G/TOP$ with corresponding 4-dimensional normal map $(f,b) : M \longrightarrow S^4$

$$\begin{aligned} (\eta,t) \;&=\; \sigma_*(f,b) \;=\; \tfrac{1}{8}\sigma(M) \;=\; -\tfrac{1}{24}p_1(\eta) \\ &\in \pi_4(G/TOP) \;=\; L_4(\mathbb{Z}) \;=\; \mathbb{Z}\ , \end{aligned}$$

by the Hirzebruch signature theorem. In particular, the generator $1 \in \pi_4(G/TOP) = \mathbb{Z}$ is represented by a fibre homotopy trivialized topological bundle $\eta : S^4 \longrightarrow BTOP$ such that

$$\begin{aligned} p_1(\eta) \;&=\; -24 \in H^4(S^4) \;=\; \mathbb{Z}\ , \\ \kappa(\eta) \;&=\; 1 \in H^4(S^4;\mathbb{Z}_2). \end{aligned}$$

This corresponds to a normal map $(f,b) : M \longrightarrow S^4$ where $M$ is the 4-dimensional Freedman $E_8$-manifold. For any map $(\widetilde{\eta},t) : S^4 \longrightarrow G/PL$ and the corresponding 4-dimensional $PL$ normal map $(f,\widetilde{b}) : M \longrightarrow S^4$

$$\begin{aligned} &p_1(\widetilde{\eta}) \equiv 0(\bmod 48)\ , \\ &\sigma(M) \;=\; -\tfrac{1}{3}p_1(\widetilde{\eta}) \equiv 0(\bmod 16) \end{aligned}$$

by Rochlin's theorem, so that

$$\begin{aligned} (\widetilde{\eta},t) \;&=\; \sigma_*(f,\widetilde{b}) \;=\; \tfrac{1}{8}\sigma(M) \;=\; -\tfrac{1}{24}p_1(\widetilde{\eta}) \\ &\in \pi_4(G/PL) \;=\; 2\mathbb{Z} \subset \pi_4(G/TOP) \;=\; \mathbb{Z}\ . \end{aligned}$$

The natural map $G/PL \longrightarrow G/TOP$ induces isomorphisms

$$\pi_m(G/PL) \xrightarrow{\cong} \pi_m(G/TOP) = L_m(\mathbb{Z}) \ (m \neq 4)$$

and multiplication by 2 in dimension 4

$$2 : \pi_4(G/PL) = \mathbb{Z} \longrightarrow \pi_4(G/TOP) = L_4(\mathbb{Z}) = \mathbb{Z} ,$$

and there are isomorphisms

$$\begin{aligned} \pi_4(G/TOP) &\xrightarrow{\cong} \mathbb{Z} \ ; \ ((\eta, t) : S^4 \longrightarrow G/TOP) \longmapsto \tfrac{1}{24} p_1(\eta) , \\ \pi_4(G/PL) &\xrightarrow{\cong} \mathbb{Z} \ ; \ ((\widetilde{\eta}, t) : S^4 \longrightarrow G/PL) \longmapsto \tfrac{1}{48} p_1(\widetilde{\eta}) . \end{aligned}$$

See Milgram [24] for a detailed comparison in dimensions $\leq 7$ of the homotopy types of $BTOP$, $BPL$ and $BG$.

By definition, the **structure set** $\mathbb{S}^{TOP}(M)$ of a closed $m$-dimensional topological manifold $M$ consists of equivalence classes of pairs $(L, f)$ where $L$ is a closed $m$-dimensional topological manifold and $f : L \longrightarrow M$ is a homotopy equivalence. Equivalence is defined by $(L, f) \sim (L', f')$ if $f'^{-1}f : L \longrightarrow L'$ is homotopic to a homeomorphism. There is a similar definition in the $PL$ category of $\mathbb{S}^{PL}(M)$. The structure sets for $m \geq 5$ (or $m = 4$ and $\pi_1(M)$ good for $TOP$) fit into a commutative braid of Sullivan-Wall surgery exact sequences of pointed sets

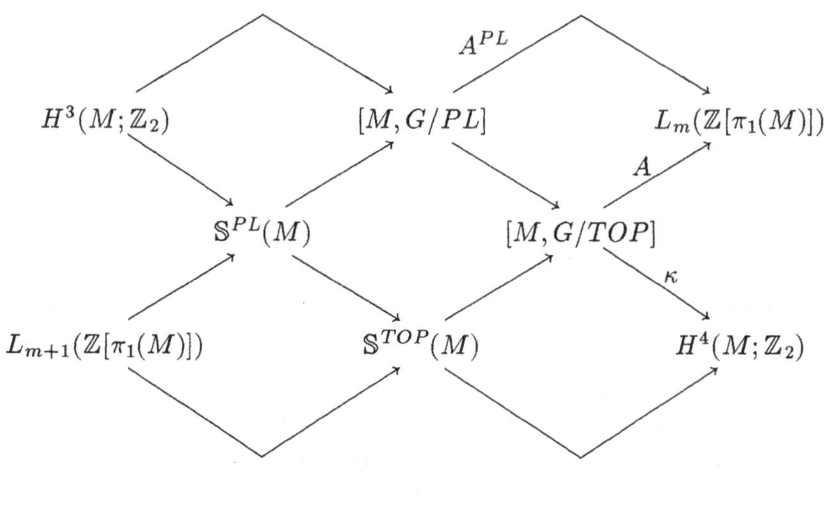

where $L_*(\mathbb{Z}[\pi_1(M)])$ are the surgery obstruction groups and

$$\mathbb{S}^{TOP}(M) \longrightarrow H^4(M;\mathbb{Z}_2) \; ; \; (L,f) \longmapsto (f^{-1})^*\kappa(L) - \kappa(M) \; .$$

The $TOP$ surgery exact sequence was expressed algebraically in Ranicki [33] as an exact sequence of abelian groups

$$\ldots \longrightarrow L_{m+1}(\mathbb{Z}[\pi_1(M)]) \longrightarrow \mathbb{S}^{TOP}(M) \longrightarrow H_m(M;\mathbb{L}_\cdot) \xrightarrow{A} L_m(\mathbb{Z}[\pi_1(M)])$$

where $\mathbb{L}_\cdot$ is the 1-connective quadratic $L$-spectrum such that

$$\pi_*(\mathbb{L}_\cdot) = L_*(\mathbb{Z}) \;\; (* \geq 1) \; ,$$

the generalized homology groups $H_*(M;\mathbb{L}_\cdot)$ are the cobordism groups of sheaves over $M$ of locally quadratic Poincaré complexes over $\mathbb{Z}$, and

$$A \; : \; [M, G/TOP] = H_m(M;\mathbb{L}_\cdot) \longrightarrow L_m(\mathbb{Z}[\pi_1(M)])$$

is the algebraic $L$-theory assembly map.

**Proposition.** (Siebenmann [42, §15], Hollingsworth and Morgan [14], Morita [29])
(i) *For any space $M$*

$$\begin{aligned}&\mathrm{im}(\kappa : [M, BTOP] \longrightarrow H^4(M;\mathbb{Z}_2)) \\ &\quad = \mathrm{im}((r_2 \;\; Sq^2) : H^4(M;\mathbb{Z}) \oplus H^2(M;\mathbb{Z}_2) \longrightarrow H^4(M;\mathbb{Z}_2)) \; ,\end{aligned}$$

*where $r_2$ is reduction mod 2.*
(ii) *For a closed $m$-dimensional topological manifold $M$ with $m \geq 5$, or $m = 4$ and $\pi_1(M)$ good, the image of the function $\mathbb{S}^{TOP}(M) \longrightarrow H^4(M;\mathbb{Z}_2)$ is the subgroup*

$$\mathrm{im}(\kappa : \ker(A) \longrightarrow H^4(M;\mathbb{Z}_2)) \subseteq \mathrm{im}(\kappa : [M, BTOP] \longrightarrow H^4(M;\mathbb{Z}_2)) \; ,$$

*with equality if*

$$\mathrm{im}(A) = \mathrm{im}(A^{PL} : [M, G/PL] \longrightarrow L_m(\mathbb{Z}[\pi_1(M)])) \; .$$

**Example.** (Hsiang and Shaneson [15], Wall [49, 15A], Kirby and Siebenmann [18]). The surgery classification of $PL$ structures on tori is an essential ingredient of the Kirby-Siebenmann structure theory of topological manifolds. The assembly map

$$A \; : \; H_n(T^m;\mathbb{L}_\cdot) \longrightarrow L_n(\mathbb{Z}[\mathbb{Z}^m])$$

is an isomorphism for $n \geq m+1$, and a split injection with cokernel $L_0(\mathbb{Z})$ for $n = m$. (This was first obtained geometrically by Shaneson and Wall, and then proved algebraically by Novikov and Ranicki).

The braid of surgery exact sequences of $T^m$ $(m \geq 5)$

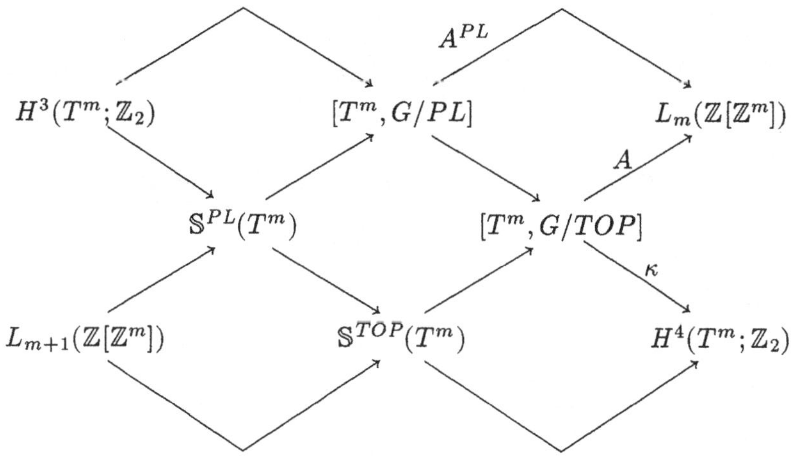


has

$$\mathbb{S}^{TOP}(T^m) \ = \ 0 \ ,$$

$$\mathbb{S}^{PL}(T^m) \ = \ [T^m, TOP/PL] \ = \ H^3(T^m;\mathbb{Z}_2) \ .$$

Thus every closed $m$-dimensional topological manifold homotopy equivalent to $T^m$ is homeomorphic to $T^m$, but does not carry a unique $PL$ structure. A **fake torus** is a closed $m$-dimensional $PL$ manifold $\tau^m$ which is homeomorphic but not $PL$ homeomorphic to $T^m$. Every element

$$\kappa \neq 0 \in \mathbb{S}^{PL}(T^m) \ = \ H^3(T^m;\mathbb{Z}_2)$$

is represented by a triangulation $(\tau^m, f)$ of $T^m$ by a fake torus $\tau^m$ such that $\kappa(f) = \kappa$. The homeomorphism $f : \tau^m \longrightarrow T^m$ is not homotopic to a $PL$ homeomorphism, constituting a counterexample to the Manifold Hauptvermutung. The application to topological manifold structure theory makes use of the fact that $f$ lifts to a homeomorphism $\overline{f} : \overline{\tau}^m \longrightarrow \overline{T}^m$ of finite covers which is homotopic to a $PL$ homeomorphism, i.e. every fake torus has a finite cover which is $PL$ homeomorphic to a genuine torus.

**Example.** The braid of surgery exact sequences of $T^m \times S^k$ $(m+k \geq 5,\ k \geq 2)$

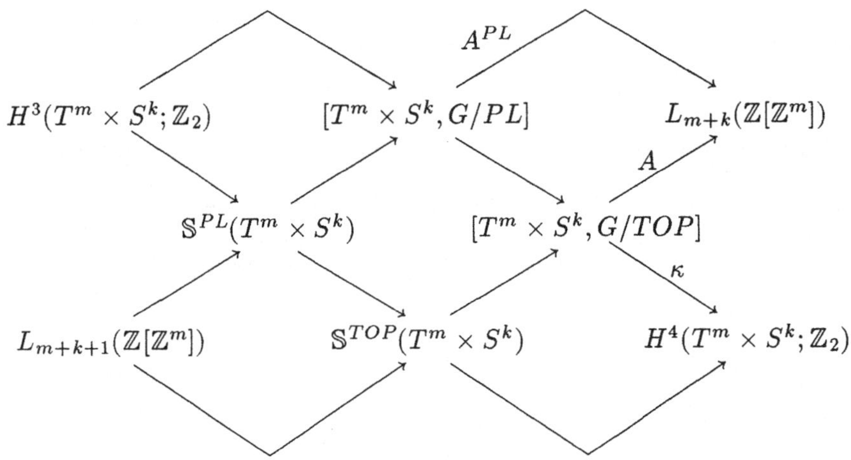


has

$$\begin{aligned} \mathbb{S}^{TOP}(T^m \times S^k) &= [T^m, G/TOP] , \\ \mathbb{S}^{PL}(T^m \times S^k) &= [T^m, G/PL] \oplus H^{3-k}(T^m; \mathbb{Z}_2) \end{aligned}$$

with

$$\mathrm{im}(\kappa : \mathbb{S}^{TOP}(T^m \times S^k) \longrightarrow H^4(T^m \times S^k; \mathbb{Z}_2)) = H^4(T^m; \mathbb{Z}_2) \subset H^4(T^m \times S^k; \mathbb{Z}_2) .$$

For every element

$$\kappa \in H^4(T^m; \mathbb{Z}_2) \subset H^4(T^m \times S^k; \mathbb{Z}_2)$$

there exists $(L, f) \in \mathbb{S}^{TOP}(T^m \times S^k)$ with $(f^*)^{-1}\kappa(L) = \kappa$. If $\kappa \neq 0$ the closed $(m+k)$-dimensional topological manifold $L$ does not admit a $PL$ structure, constituting a counterexample to the Combinatorial Triangulation Conjecture. (After Freedman $\mathbb{S}^{TOP}(T^m \times S^k) = [T^m, G/TOP]$ also in the case $m+k=4$.) See Siebenmann [42, §2], Kirby and Siebenmann [18, pp. 210-213] for explicit constructions of such high-dimensional torus-related counterexamples to the Hauptvermutung and Combinatorial Triangulation Conjecture, starting from the Milnor $E_8$-plumbing 4-dimensional $PL$ manifold $(Q^4, \Sigma^3)$ with boundary the Poincaré homology sphere $\Sigma$. See Scharlemann [38] for such explicit constructions of manifolds without combinatorial triangulation in the homotopy types of $S^3 \times S^1 \# S^2 \times S^2$, $T^2 \times S^3$ and $\mathbb{CP}^2 \times S^1$.

The original Milnor differentiable exotic spheres arose as the total spaces of $PL$ trivial differentiable sphere bundles over $S^4$. The original Novikov examples of homotopy equivalences of high-dimensional differentiable manifolds which are not homotopic to diffeomorphisms were constructed using fibre homotopy trivializations of differentiable sphere bundles over $S^4$. Likewise, fibre homotopy trivial topological sphere bundles over $S^4$ provided examples of topological manifolds without a combinatorial triangulation:

**Example.** The structure sets of $S^m \times S^n$ with $m, n \geq 2$ are such that

$$\begin{aligned}\mathbb{S}^{TOP}(S^m \times S^n) &= L_m(\mathbb{Z}) \oplus L_n(\mathbb{Z}) \text{ if } m+n \geq 4 \\ \mathbb{S}^{PL}(S^m \times S^n) &= \widetilde{L}_m(\mathbb{Z}) \oplus \widetilde{L}_n(\mathbb{Z}) \text{ if } m+n \geq 5\end{aligned}$$

where

$$\begin{aligned}\widetilde{L}_m(\mathbb{Z}) &= \pi_m(G/PL) \\ &= \begin{cases} \pi_m(G/TOP) = L_m(\mathbb{Z}) & \text{if } m \neq 4 \\ 2\pi_4(G/TOP) = 2L_4(\mathbb{Z}) & \text{if } m = 4 \end{cases}\end{aligned}$$

(Ranicki [33, 20.4]). For any element $(W, f) \in \mathbb{S}^{TOP}(S^m \times S^n)$ it is possible to make the homotopy equivalence $f : W \longrightarrow S^m \times S^n$ topologically transverse regular at $S^m \times \{*\}$ and $\{*\} \times S^n \subset S^m \times S^n$. The restrictions of $f$ are normal maps

$$\begin{aligned}(f_M, b_M) &= f| : M^m = f^{-1}(S^m \times \{*\}) \longrightarrow S^m , \\ (f_N, b_N) &= f| : N^n = f^{-1}(\{*\} \times S^n) \longrightarrow S^n\end{aligned}$$

such that

$$(W, f) = (\sigma_*(f_M, b_M), \sigma_*(f_N, b_N)) \in \mathbb{S}^{TOP}(S^m \times S^n) = L_m(\mathbb{Z}) \oplus L_n(\mathbb{Z}) .$$

Every element

$$\begin{aligned}x \in L_m(\mathbb{Z}) &= \pi_m(G/TOP) \\ &= \pi_{m+1}(B\widetilde{TOP}(n+1) \longrightarrow BG(n+1)) \quad (n \geq 2)\end{aligned}$$

is realized by a topological block bundle

$$\eta : S^m \longrightarrow B\widetilde{TOP}(n+1)$$

with a fibre homotopy trivial topological sphere bundle

$$S^n \longrightarrow S(\eta) \longrightarrow S^m .$$

Making the degree 1 map $\rho : S^{m+n} \longrightarrow T(\eta)$ topologically transverse regular at $S^m \subset T(\eta)$ gives an $m$-dimensional normal map

$$(f_M, b_M) = \rho| : M^m = \rho^{-1}(S^m) \longrightarrow S^m$$

with $b_M : \nu_M \longrightarrow \eta$, such that the surgery obstruction is

$$\sigma_*(f_M, b_M) = x \in L_m(\mathbb{Z}) .$$

The closed $(m+n)$-dimensional topological manifold $S(\eta)$ is equipped with a homotopy equivalence $f : S(\eta) \longrightarrow S^m \times S^n$ such that

$$(S(\eta), f) = (x, 0) \in \mathbb{S}^{TOP}(S^m \times S^n) = L_m(\mathbb{Z}) \oplus L_n(\mathbb{Z}) ,$$

where $f^{-1}(S^m \times \{*\}) = M$. The normal bundle of $S(\eta)$ is classified by

$$\nu_{S(\eta)} : S(\eta) \xrightarrow{f} S^m \times S^n \xrightarrow{proj.} S^m \xrightarrow{-\eta} BTOP ,$$

with the Kirby-Siebenmann invariant given by

$$\kappa(S(\eta)) \;=\; \kappa(\eta) \;=\; \begin{cases} x(\text{mod}2) & \text{if } m=4 \\ 0 & \text{if } m\neq 4 \end{cases}$$
$$\in \operatorname{im}(H^4(S^m;\mathbb{Z}_2)\longrightarrow H^4(S(\eta);\mathbb{Z}_2))$$

where

$$\begin{aligned} H^4(S^m;\mathbb{Z}_2) &= \operatorname{coker}(\widetilde{L}_m(\mathbb{Z})\longrightarrow L_m(\mathbb{Z})) \\ &= \pi_{m-1}(TOP/PL) \;=\; \begin{cases} \mathbb{Z}_2 & \text{if } m=4 \\ 0 & \text{if } m\neq 4 \,. \end{cases} \end{aligned}$$

The surgery classifying space $G/TOP$ fits into a fibration sequence

$$G/TOP \longrightarrow B\widetilde{TOP}(n) \longrightarrow BG(n)$$

for any $n\geq 3$, by a result of Rourke and Sanderson [34]. The generator

$$\begin{aligned} 1\in L_4(\mathbb{Z}) &= \pi_5(B\widetilde{TOP}(3)\longrightarrow BG(3)) \\ &= \pi_4(G/TOP) \;=\; \mathbb{Z} \end{aligned}$$

is represented by a map $(\eta,t)\,:\,S^4\longrightarrow G/TOP$ corresponding to a topological block bundle $\eta\,:\,S^4\longrightarrow B\widetilde{TOP}(3)$ with a fibre homotopy trivialization $t\,:\,\eta\simeq\{*\}:S^4\longrightarrow BG(3)$, such that

$$\begin{aligned} p_1(\eta) &= -24\in H^4(S^4;\mathbb{Z}) \;=\; \mathbb{Z}\,, \\ \kappa(\eta) &= 1\in H^4(S^4;\mathbb{Z}_2) \;=\; \mathbb{Z}_2\,. \end{aligned}$$

For every $n\geq 2$ the closed $(n+4)$-dimensional topological manifold

$$W^{n+4} \;=\; S(\eta\oplus\epsilon^{n-2})$$

is the total space of a fibre homotopy trivial non-$PL$ topological sphere bundle $\eta\oplus\epsilon^{n-2}$

$$S^n \longrightarrow W \longrightarrow S^4\,,$$

with a homotopy equivalence $f:W\longrightarrow S^4\times S^n$. The element

$$(W,f) \;=\; (1,0)\neq(0,0)\in\mathbb{S}^{TOP}(S^4\times S^n) \;=\; L_4(\mathbb{Z})\oplus L_n(\mathbb{Z})$$

realizes the generator

$$\begin{aligned} x \;=\; 1\in L_4(\mathbb{Z}) &= \pi_5(B\widetilde{TOP}(n+1)\longrightarrow BG(n+1)) \\ &= \pi_4(G/TOP) \;=\; \mathbb{Z}\,. \end{aligned}$$

The manifold $W$ does not admit a combinatorial triangulation, with

$$\kappa(W) \;=\; 1\in H^4(W;\mathbb{Z}_2) \;=\; \mathbb{Z}_2$$

and $M^4=f^{-1}(S^4\times\{*\})\subset W$ the 4-dimensional Freedman $E_8$-manifold.

## §5. Homology Manifolds

An $m$-**dimensional homology manifold** is a space $X$ such that the local homology groups at each $x \in X$ are given by

$$\begin{aligned} H_r(X, X\backslash\{x\}) &= H_r(\mathbb{R}^m, \mathbb{R}^m\backslash\{0\}) \\ &= \begin{cases} \mathbb{Z} & \text{if } r = m \\ 0 & \text{otherwise.} \end{cases} \end{aligned}$$

An $m$-dimensional topological manifold is an $m$-dimensional homology manifold.

The local homology groups of the polyhedron $|K|$ of a simplicial complex $K$ at $x \in |K|$ are such that

$$H_*(|K|, |K|\backslash\{x\}) = \widetilde{H}_{*-|\sigma|-1}(\mathrm{link}_K(\sigma))$$

if $x \in \mathrm{int}(\sigma)$ for a simplex $\sigma \in K$.

An $m$-**dimensional combinatorial homology manifold** is a simplicial complex $K$ such that for each $\sigma \in K$

$$H_*(\mathrm{link}_K(\sigma)) = H_*(S^{m-|\sigma|-1}) .$$

Similarly for a **combinatorial homotopy manifold**.

A $PL$ manifold is a combinatorial homotopy manifold. A combinatorial homotopy manifold is a combinatorial homology manifold. The polyhedron of a simplicial complex $K$ is an $m$-dimensional homology manifold $|K|$ if and only if $K$ is an $m$-dimensional combinatorial homology manifold.

**Example.** For $m \geq 5$ the double suspension of any $(m-2)$-dimensional combinatorial homology sphere $\Sigma$ is an $m$-dimensional combinatorial homology manifold $K$ such that the polyhedron $|K|$ is a topological manifold homeomorphic to $S^m$ (Edwards, Cannon). The combinatorial homology manifold $K$ is a combinatorial homotopy manifold if and only if $\Sigma$ is simply-connected.

More generally :

**Theorem.** (Edwards [8]) *For $m \geq 5$ the polyhedron of an $m$-dimensional combinatorial homology manifold $K$ is an $m$-dimensional topological manifold $|K|$ if and only if* $\mathrm{link}_K(\sigma)$ *is simply-connected for each vertex $\sigma \in K$.*

This includes as a special case the result of Siebenmann [41] that for $m \geq 5$ the polyhedron of an $m$-dimensional combinatorial homotopy manifold $K$ is an $m$-dimensional topological manifold $|K|$.

**Triangulation Conjecture.** *Every compact $m$-dimensional topological manifold can be triangulated by a combinatorial homology manifold.*

The triangulation need not be combinatorial, i.e. the combinatorial homology manifold need not be a $PL$ manifold.

It follows from the properties of Casson's invariant for oriented homology 3-spheres that the 4-dimensional Freedman $E_8$-manifold cannot be triangulated (Akbulut and McCarthy [1, p.xvi]), so that:

**Theorem.** *The Triangulation Conjecture is false for $m = 4$.*

The Triangulation Conjecture is unresolved for $m \geq 5$. The Kirby-Siebenmann examples of topological manifolds without a combinatorial triangulation are triangulable.

**Definition.** A **manifold homology resolution** $(M, f)$ of a space $X$ is a compact $m$-dimensional topological manifold $M$ with a surjective map $f : M \longrightarrow X$ such that the point inverses $f^{-1}(x)$ $(x \in X)$ are acyclic. Similarly for **manifold homotopy resolution**, with contractible point inverses.

A space $X$ which admits an $m$-dimensional manifold homology resolution is an $m$-dimensional homology manifold. Bryant, Ferry, Mio and Weinberger [4] have constructed compact $ANR$ homology manifolds in dimensions $m \geq 5$ which do not admit a manifold homotopy resolution.

Let $\theta_3^H$ (resp. $\theta_3^h$) be the Kervaire-Milnor cobordism group of oriented 3-dimensional combinatorial homology (resp. homotopy) spheres modulo those which bound acyclic (resp. contractible) 4-dimensional $PL$ manifolds, with addition given by connected sum.

Given a finite $m$-dimensional combinatorial homology manifold $K$ define

$$c^H(K) = \sum_{\sigma \in K^{(m-4)}} [\mathrm{link}_K(\sigma)]\sigma \in H_{m-4}(K; \theta_3^H) = H^4(K; \theta_3^H) .$$

Similarly, given a finite $m$-dimensional combinatorial homotopy manifold $K$ define

$$c^h(K) = \sum_{\sigma \in K^{(m-4)}} [\mathrm{link}_K(\sigma)]\sigma \in H_{m-4}(K; \theta_3^h) = H^4(K; \theta_3^h) .$$

**Theorem.** (Cohen [6], Sato [37], Sullivan [47, pp. 63–65])
(i) *An $m$-dimensional combinatorial homology manifold $K$ is such that*

$$c^H(K) = 0 \in H^4(K; \theta_3^H)$$

*if (and for $m \geq 5$ only if) $K$ has a $PL$ manifold homology resolution.*
(ii) *An $m$-dimensional combinatorial homotopy manifold $K$ is such that*

$$c^h(K) = 0 \in H^4(K; \theta_3^h)$$

*if (and for $m \geq 5$ only if) $K$ has a PL manifold homotopy resolution.*

The natural map $\theta_3^h \longrightarrow \theta_3^H$ is such that for a finite $m$-dimensional combinatorial homotopy manifold $K$

$$H_{m-4}(K;\theta_3^h) \longrightarrow H_{m-4}(K;\theta_3^H) \; ; \; c^h(K) \longmapsto c^H(K) \; .$$

Every oriented 3-dimensional combinatorial homology sphere $\Sigma$ bounds a parallelizable 4-dimensional $PL$ manifold $W$, allowing the Rochlin invariant of $\Sigma$ to be defined by

$$\alpha(\Sigma) \; = \; \sigma(W) \in 8\mathbb{Z}/16\mathbb{Z} \; = \; \mathbb{Z}_2$$

as in §3 above. The Rochlin invariant defines a surjection

$$\alpha \; : \; \theta_3^H \longrightarrow \mathbb{Z}_2 \; ; \; \Sigma \longmapsto \alpha(\Sigma) \; ,$$

with $\alpha(\Sigma) = 1 \in \mathbb{Z}_2$ for the Poincaré homology 3-sphere $\Sigma$.

**Remarks.** (i) Fintushel and Stern [9] applied Donaldson theory to show that the kernel of $\alpha : \theta_3^H \longrightarrow \mathbb{Z}_2$ is infinitely generated.
(ii) The composite

$$\theta_3^h \longrightarrow \theta_3^H \xrightarrow{\alpha} \mathbb{Z}_2$$

is 0, by a result of Casson (Akbulut and McCarthy [1, p.xv]).

The exact sequence of coefficient groups

$$0 \longrightarrow \ker(\alpha) \longrightarrow \theta_3^H \xrightarrow{\alpha} \mathbb{Z}_2 \longrightarrow 0$$

induces a cohomology exact sequence

$$\begin{aligned} \ldots \longrightarrow H^n(M;\ker(\alpha)) \longrightarrow H^n(M;\theta_3^H) \xrightarrow{\alpha} & \; H^n(M;\mathbb{Z}_2) \\ \xrightarrow{\delta} & \; H^{n+1}(M;\ker(\alpha)) \longrightarrow \ldots \end{aligned}$$

for any space $M$.

**Theorem.** (Galewski-Stern [11], [12], Matumoto [23])
(i) *The Kirby-Siebenmann invariant $\kappa(M) \in H^4(M;\mathbb{Z}_2)$ of a compact $m$-dimensional topological manifold $M$ is such that*

$$\delta\kappa(M) \; = \; 0 \in H^5(M;\ker(\alpha))$$

*if (and for $m \geq 5$ only if) $M$ is triangulable. If $M$ is triangulable then for any triangulation $(K, f : |K| \longrightarrow M)$*

$$\begin{aligned} \kappa(M) \; &= \; f_*\alpha(c^H(K)) \in \mathrm{im}(\alpha : H_{m-4}(M;\theta_3^H) \longrightarrow H_{m-4}(M;\mathbb{Z}_2)) \\ &= \; \mathrm{im}(\alpha : H^4(M;\theta_3^H) \longrightarrow H^4(M;\mathbb{Z}_2)) \\ &= \; \ker(\delta : H^4(M;\mathbb{Z}_2) \longrightarrow H^5(M;\ker(\alpha))) \; . \end{aligned}$$

(ii) *Every finite m-dimensional combinatorial homology manifold K for $m \geq 5$ admits a topological manifold homotopy resolution $(M, f : |K| \longrightarrow M)$ such that M is a triangulable m-dimensional topological manifold with*

$$\begin{aligned}\kappa(M) &= f_*\alpha(c^H(K)) \in \mathrm{im}(\alpha : H_{m-4}(M;\theta_3^H) \longrightarrow H_{m-4}(M;\mathbb{Z}_2)) \\ &= \mathrm{im}(\alpha : H^4(M;\theta_3^H) \longrightarrow H^4(M;\mathbb{Z}_2)) .\end{aligned}$$

(iii) *The Triangulation Conjecture is true for every $m \geq 5$ if and only if the surjection $\alpha : \theta_3^H \longrightarrow \mathbb{Z}_2$ splits, i.e. if and only if there exists a 3-dimensional combinatorial homology sphere $\Sigma$ such that $\alpha(\Sigma) = 1 \in \mathbb{Z}_2$ and $2(\Sigma) = 0 \in \theta_3^H$.*
(iv) *The stable classifying spaces $BPL, BTOP, BH$ for the bundle theories associated to PL, topological and combinatorial homology manifolds are related by a braid of fibrations*

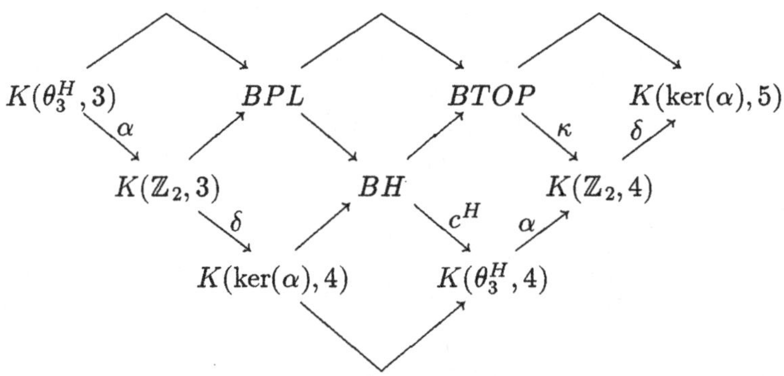

The Cohen-Sato-Sullivan $PL$ manifold homology resolution obstruction $c^H(K) \in H^4(M;\theta_3^H)$ of an $m$-dimensional combinatorial homology manifold $K$ is the homotopy class of the composite

$$K \xrightarrow{\nu_K} BH \xrightarrow{c^H} K(\theta_3^H, 4) .$$

The Galewski-Matumoto-Stern triangulation obstruction $\delta\kappa(M) \in H^5(M;\ker(\alpha))$ of an $m$-dimensional topological manifold $M$ is the homotopy class of the composite

$$M \xrightarrow{\nu_M} BTOP \longrightarrow K(\ker(\alpha), 5) .$$

**Example.** Let $(Q^4, \Sigma)$ be the Milnor 4-dimensional $PL$ $E_8$-manifold with boundary the Poincaré 3-dimensional homology sphere, such that $\sigma(Q) = 8$, $\kappa(Q) = 0$, $\alpha(\Sigma) = 1$. Coning off the boundary of $Q$ defines a 4-dimensional combinatorial homology manifold $K = Q \cup_\Sigma c\Sigma$ such that

$$c^H(K) = [\Sigma] \neq 0 \in H^4(K;\theta_3^H) = \theta_3^H ,$$

so that $K$ does not have a $PL$ manifold homology resolution. Let $(W, \Sigma)$ be the contractible Freedman 4-dimensional topological $E_8$-manifold, such that $\sigma(W) =$

8, $\kappa(W) = 1$ (as at the end of §3). The polyhedron $|K|$ admits a manifold homotopy resolution $f : M = Q \cup_\Sigma W \longrightarrow |K|$, with the non-triangulable closed 4-dimensional topological manifold $M$ such that

$$\kappa(M) = [f^* c^H(K)] \neq 0 \in \operatorname{im}(\alpha : H^4(M;\theta_3^H) \longrightarrow H^4(M;\mathbb{Z}_2)) .$$

The product $|K| \times S^1$ is a 5-dimensional topological manifold, with $f \times 1 : M \times S^1 \longrightarrow |K| \times S^1$ homotopic to a homeomorphism triangulating $M \times S^1$ by $K \times S^1$. The Kirby-Siebenmann invariant of $M \times S^1$ is

$$\kappa(M \times S^1) = p^*\kappa(M) \neq 0 \in \operatorname{im}(\alpha : H^4(M \times S^1;\theta_3^H) \longrightarrow H^4(M \times S^1;\mathbb{Z}_2))$$

with $p : M \times S^1 \longrightarrow M$ the projection, so that $M \times S^1$ is a triangulable 5-dimensional topological manifold without a combinatorial triangulation. In fact, $M \times S^1$ is not even homotopy equivalent to a 5-dimensional $PL$ manifold.

The rel $\partial$ version of the Cohen-Sato-Sullivan $PL$ manifold resolution obstruction theory applies to the problem of deforming a $PL$ map of $PL$ manifolds with acyclic (resp. contractible) point inverses to a $PL$ homeomorphism, and the rel $\partial$ version of the Galewski-Matumoto-Stern triangulation obstruction theory applies to the problem of deforming a homeomorphism of $PL$ manifolds to a $PL$ map with acyclic point inverses, as follows.

Let $f : K \longrightarrow L$ be a $PL$ map of compact $m$ dimensional $PL$ manifolds, with acyclic (resp. contractible) point inverses $f^{-1}(x)$ $(x \in L)$. The mapping cylinder $W = K \times I \cup_f L$ is an $(m+1)$-dimensional combinatorial homology (resp. homotopy) manifold with $PL$ manifold boundary and a $PL$ map

$$(g;f,1) \; : \; (W;K,L) \longrightarrow L \times (I;\{0\},\{1\})$$

with acyclic (resp. contractible) point inverses. For each simplex $\sigma \in L$ let $D(\sigma, L)$ be the dual cell in the barycentric subdivision $L'$ of $L$, such that there is a $PL$ homeomorphism

$$(D(\sigma,L),\partial D(\sigma,L)) \cong (D^{m-|\sigma|},S^{m-|\sigma|-1}) .$$

The combinatorial $(m+1-|\sigma|)$-dimensional homology (resp. homotopy) manifold

$$(W_\sigma,\partial W_\sigma) = g^{-1}(D(\sigma,L) \times I, \partial(D(\sigma,L) \times I))$$

is such that the restriction

$$g| \; : \; (W_\sigma,\partial W_\sigma) \longrightarrow (D(\sigma,L) \times I, \partial(D(\sigma,L) \times I)) \cong (D^{m+1-|\sigma|},S^{m-|\sigma|})$$

is a homology (resp. homotopy) equivalence, with

$$\partial W_\sigma = f^{-1}D(\sigma,L) \cup g^{-1}(\partial D(\sigma,L) \times I) \cup D(\sigma,L) .$$

If $f$ has acyclic point inverses define

$$\begin{aligned} c^H(f) &= c^H_\partial(W;K,L) \\ &= \sum_{\sigma \in L^{(m-3)}} [\partial W_\sigma]\sigma \in H_{m-3}(L;\theta_3^H) = H^3(L;\theta_3^H) , \end{aligned}$$

and if $f$ has contractible point inverses define

$$\begin{aligned} c^h(f) &= c^h_\partial(W;K,L) \\ &= \sum_{\sigma\in L^{(m-3)}} [\partial W_\sigma]\sigma \in H_{m-3}(L;\theta^h_3) = H^3(L;\theta^h_3) . \end{aligned}$$

**Proposition.** *A PL map of compact m-dimensional PL manifolds $f : K \longrightarrow L$ with acyclic (resp. contractible) point inverses is such that $c^H(f) = 0$ (resp. $c^h(f) = 0$) if (and for $m \geq 5$ only if) $f$ is concordant to a PL homeomorphism, i.e. homotopic to a PL homeomorphism through PL maps with acyclic (resp. contractible) point inverses.*

**Remark.** Cohen [6] actually proved that for $m \geq 5$ a $PL$ map $f : K \longrightarrow L$ of $m$-dimensional combinatorial homotopy manifolds with contractible point inverses is homotopic through $PL$ maps with contractible point inverses to a $PL$ map $F : K \longrightarrow L$ which is a homeomorphism. If $K, L$ are $PL$ manifolds then $F$ can be chosen to be a $PL$ homeomorphism, so that $c^h(f) = 0$.

Returning to the Manifold Hauptvermutung, we have:

**Proposition.** *Let $f : |K| \longrightarrow |L|$ be a homeomorphism of the polyhedra of compact m-dimensional PL manifolds $K, L$. The Casson-Sullivan invariant $\kappa(f) \in H^3(L;\mathbb{Z}_2)$ is such that*

$$\delta\kappa(f) = 0 \in H^4(L;\ker(\alpha))$$

*if (and for $m \geq 5$ only) if $f$ is homotopic to a PL map $F : K \longrightarrow L$ with acyclic point inverses, in which case $\kappa(f)$ is the image under $\alpha$ of the Cohen-Sato-Sullivan invariant $c^H(F)$*

$$\begin{aligned} \kappa(f) &= \alpha(c^H(F)) \in \mathrm{im}(\alpha : H^3(L;\theta^H_3) \longrightarrow H^3(L;\mathbb{Z}_2)) \\ &= \ker(\delta : H^3(L;\mathbb{Z}_2) \longrightarrow H^4(L;\ker(\alpha))) . \end{aligned}$$

**Proof.** The mapping cylinder $W = |K| \times I \cup_f |L|$ of $f$ is an $(m+1)$-dimensional topological manifold with $PL$ boundary $\partial W = |K| \times \{0\} \cup |L|$, and with a homeomorphism

$$(g; f, 1) : (W; |K|, |L|) \longrightarrow |L| \times (I; \{0\}, \{1\}) .$$

By the rel $\partial$ version of the Galewski-Matumoto-Stern obstruction theory $\kappa(f) \in \mathrm{im}(\alpha)$ if (and for $m \geq 5$ only if) the triangulation of $\partial W$ extends to a triangulation of $W$, in which case it is possible to approximate $g$ by a $PL$ map $G$ such that the restriction $G| = F : K \longrightarrow L$ is a $PL$ map homotopic to $f$ with acyclic point inverses and $\kappa(f) = \alpha(c^H(F))$.

**Corollary.** *If the Triangulation Conjecture is true for every $m \geq 5$ (i.e. if $\alpha : \theta^H_3 \longrightarrow \mathbb{Z}_2$ splits) every homeomorphism of compact m-dimensional PL manifolds* *is homotopic to a PL map with acyclic point inverses.*

## References


[1] S. Akbulut and J. D. McCarthy, **Casson's invariant for oriented homology 3-spheres.** Mathematical Notes 36, Princeton (1990)

[2] P. Alexandroff and H. Hopf, **Topologie.** Springer (1935) Chelsea (1972)

[3] M. A. Armstrong, G. E. Cooke and C. P. Rourke, **The Princeton notes on the Hauptvermutung.** Princeton (1968), Warwick (1972) (Published in this volume)

[4] J. Bryant, S. Ferry, W. Mio and S. Weinberger, **The topology of homology manifolds.** Bull. A.M.S. 28, 324–328 (1993)

[5] A. J. Casson, **Generalizations and applications of block bundles.** Fellowship dissertation, Trinity College Library, Cambridge (1967) (Published in this volume)

[6] M. Cohen, **Homeomorphisms between homotopy manifolds and their resolutions.** Inv. Math. 10, 239–250 (1970)

[7] –, **A course in simple homotopy theory.** Graduate Texts in Mathematics 10, Springer (1973)

[8] R. D. Edwards, **The topology of manifolds and cell-like maps.** Proc. 1978 I. C. M. Helsinki, 111–127 (1980)

[9] R. Fintushel and R. J. Stern, **Instanton homology of Seifert fibred homology three spheres.** Proc. London. Math. Soc. 61, 109–137 (1990)

[10] M. Freedman and F. Quinn, **The topology of 4-manifolds.** Princeton (1990)

[11] D. E. Galewski and R. J. Stern, **The relationship between homology and topological manifolds via homology transversality.** Inv. Math. 39, 277–292 (1977)

[12] –, **Classification of simplicial triangulations of topological manifolds.** Ann. of Maths. 111, 1–34 (1980)

[13] L. Guillou and A. Marin, **A la Recherche de la Topologie Perdue.** Progress in Math. 62, Birkhäuser (1986)

[14] J. Hollingsworth and J. Morgan, **Homotopy triangulations and topological manifolds.** Preprint (1970)

[15] W. C. Hsiang and J. Shaneson, **Fake tori.** Topology of Manifolds, Proc. 1969 Georgia Topology Conference, Markham Press, 18–51 (1970)

[16] B. Hughes and A. A. Ranicki, **Ends of complexes.** Tracts in Mathematics, Cambridge (to appear)

[17] R. Kirby, **The Topology of 4-manifolds.** Springer Lecture Notes 1374 (1989)

[18] – and L. Siebenmann, **On the triangulation of manifolds and the Hauptvermutung.** Bull. Amer. Math. Soc. 75, 742–749 (1969) (also in [19] 299–306)

[19] –, **Foundational essays on topological manifolds smoothings and**

**triangulations.** Ann. Math. Stud. 88, Princeton (1977)
[20] H. Kneser, **Die Topologie der Mannigfaltigkeiten.** Jahresbericht der D. M. V. 34, 1–14 (1925)
[21] R. K. Lashof and M. G. Rothenberg, **Hauptvermutung for manifolds.** Proc. 1967 Conference on the Topology of Manifolds, Prindle, Weber and Schmidt, 81–105 (1968)
[22] I. Madsen and R. J. Milgram, **The classifying spaces for surgery and cobordism of manifolds.** Ann. Math. Study 92, Princeton (1979)
[23] T. Matumoto, **Triangulations of manifolds.** A. M. S. Proc. Symposia in Pure Math. 32, 3–6 (1978)
[24] R. J. Milgram, **Some remarks on the Kirby-Siebenmann class.** Proc. 1987 Göttingen Conference on Algebraic Topology and Transformation Groups, Springer Lecture Notes 1361, 247–252 (1988)
[25] J. Milnor, **Two complexes which are homeomorphic but combinatorially distinct.** Ann. of Maths. 74, 575–590 (1961)
[26] –, **Whitehead torsion.** Bull. A. M. S. 72, 358–426 (1966)
[27] – and D. Husemoller, **Symmetric bilinear forms.** Ergebnisse der Mathematik 73, Springer (1973)
[28] E. Moïse, **Affine structures in 3-manifolds.** Ann. of Maths. 58, 458–480 (1953)
[29] S. Morita, **Some remarks on the Kirby-Siebenmann class**. J. Fac. Sci. Univ. Tokyo, Sect. I, 18, 155–161 (1972)
[30] C. Morlet, **Hauptvermutung et triangulation des variétés.** Séminaire Bourbaki 362 (1969)
[31] S. P. Novikov, **On manifolds with free abelian fundamental group and applications (Pontrjagin classes, smoothings, high-dimensional knots).** Izv. Akad. Nauk SSSR, ser. mat. 30, 208–246 (1966)
[32] Ch. Papakyriakopoulos, **A new proof of the invariance of the homology groups of a complex.** Bull. Soc. Math. Grèce 22, 1–154 (1943)
[33] A. A. Ranicki, **Algebraic $L$-theory and topological manifolds.** Tracts in Mathematics 102, Cambridge (1992)
[34] C. P. Rourke and B. J. Sanderson, **On topological neighbourhoods.** Compositio Math. 22, 387–424 (1970)
[35] – and –, **Piecewise linear topology.** Ergebnisse der Mathematik 69, Springer (1972).
[36] Y. B. Rudyak, **Piecewise-linear structures on topological manifolds.** Appendix to the Russian translation of Madsen and Milgram [22], Mir (1984)
[37] H. Sato, **Constructing manifolds by homotopy equivalences I. An obstruction to constructing $PL$ manifolds from homology manifolds.** Ann. Inst. Fourier, Grenoble 22, 271–286 (1972)
[38] M. Scharlemann, **Constructing strange manifolds with the dodecahedral space.** Duke Math. J. 43, 33–40 (1976)
[39] H. Seifert and W. Threlfall, **Lehrbuch der Topologie.** Teubner (1934).

English translation: **A textbook of topology.** Academic Press (1980)

[40] L. Siebenmann, **The obstruction to finding a boundary for an open manifold of dimension greater than five.** Princeton Ph.D. Thesis (1966)

[41] –, **Are non-triangulable manifolds triangulable?** Topology of Manifolds, Proc. 1969 Georgia Topology Conference, Markham Press, 77–84 (1970)

[42] –, **Topological manifolds.** Proc. 1970 Nice I. C. M. Vol. 2, 133–163 (1971) (also in [19] 307–337)

[43] J. Stallings, **On infinite processes leading to differentiability in the complement of a point.** Differential and combinatorial topology. (A symposium in honor of M. Morse). Princeton, 245–254 (1965)

[44] E. Steinitz, **Beiträge zur Analysis situs.** Sitz.-Ber. Berl. Math. Ges. 7, 29–49 (1908)

[45] D. P. Sullivan, **Triangulating and smoothing homotopy equivalences and homeomorphisms. Geometric Topology Seminar Notes.** Princeton (1967) (Published in this volume)

[46] –, **On the Hauptvermutung for manifolds.** Bull. Amer. Math. Soc. 73, 598–600 (1967)

[47] –, **Geometric periodicity and the invariants of manifolds.** Proc. 1970 Amsterdam Conference on Manifolds, Springer Lecture Notes in Mathematics 197, 44–75 (1971)

[48] H. Tietze, **Über die topologischen Invarianten mehrdimensionaler Mannigfaltigkeiten.** Monatsh. für Math. und Phys. 19, 1–118 (1908)

[49] C. T. C. Wall, **Surgery on compact manifolds.** Academic Press (1970)

University of Edinburgh
January 1996

# Generalisations and applications of block bundles

By A. J. Casson

## Introduction

Rourke and Sanderson [10] introduced the idea of a **block bundle**. They used block bundles in the $PL$ (piecewise linear) category as a substitute for the normal bundles of differential topology. Their block bundles had fibre $I^q$ (the unit cube in $q$-dimensional space).

We generalise the idea to allow any compact $PL$ manifold $F$ as fibre. Chapter I sets up the theory; in particular, it is shown that there is a classifying space $B\widetilde{PL}_F$ for block bundles with fibre $F$.

In Chapter II we compare block bundles with Hurewicz fibrations. Let $F$ be a compact $PL$ manifold with boundary $\partial F$, and let $BG_F$ classify Hurewicz fibrations with fibre $(F, \partial F)$. We produce a map $\chi : B\widetilde{PL}_F \longrightarrow BG_F$, arising from a natural transformation of bundle functors.

We wish to obtain information about $B\widetilde{PL}_F$; in fact we can study $BG_F$ (which is purely homotopy theoretic) and the fibre $G_F/\widetilde{PL}_F$ of $\chi$. In Chapter III we construct a map $\theta : G_F/\widetilde{PL}_F \longrightarrow (G/PL)^F$, where $G/PL$ is the space studied in Sullivan's thesis under the name $F/PL$, and $(G/PL)^F$ is the space of all unbased maps from $F$ to $G/PL$. Theorems 5,6 show that, under suitable conditions, $\theta$ is almost a homotopy equivalence. For these results it is essential to work with block bundles rather than fibre bundles.

Sullivan shows in his thesis (see [13] for a summary) that $G/PL$ is closely related to the problem of classifying $PL$ manifolds homotopy equivalent to a given manifold. Therefore it is important to have information about the homotopy type of $G/PL$. In Chapter IV we apply Theorem 5 to show that $\Omega^4(G/PL)$ is homotopy equivalent to $\Omega^8(G/PL)$; it is almost true that $G/PL$ is homotopy equivalent to $\Omega^4(G/PL)$.

In Chapter V (which is almost independent of the earlier chapters) we show that, with certain restrictions on the base-space, a block bundle with fibre $\mathbb{R}^q$ which is topologically trivial is necessarily piecewise linearly trivial. It follows from this (again using the results of [13]) that the Hauptvermutung is true for closed 1-connected $PL$ manifolds $M$ with $\dim M \geq 5$ and $H^3(M; \mathbb{Z}_2) = 0$.

I should like to thank Professor C.T.C.Wall for suggesting the study of generalized block bundles and for much encouragement. I am also very grateful to Dr. D.P.Sullivan for several conversations during the summer of 1966.

**Origins of the ideas**

Chapter I is based on §1 of [10]; the definitions and technical details are new, but the general plan is similar. The use of block bundles with arbitrary fibres was suggested to me by Professor Wall.

Chapter II is mainly technical, and new as far as I know.

Chapter III generalizes results in Sullivan's thesis (but the proofs are based on the references given rather than on Sullivan's work).

The result and method of proof in Chapter IV is new, as far as I know.

I believe that Sullivan* has a stronger result than Theorem 8 of Chapter V, but have not seen his proofs. I proved Theorem 8 before hearing of Sullivan's latest result. My proof is an extension of the idea of [16].

Summer, 1967

---

* See D.P.Sullivan, **On the Hauptvermutung for manifolds**, Bull. Amer. Math. Soc. 73 (1967) 598-600. The theorem announced there includes ours, but the proof seems somewhat different.

## I. Block Bundles

A **polyhedron** is a topological space together with a maximal family of $PL$ related locally finite triangulations. A **cell complex** $B$ is a collection of cells $PL$ embedded in a polyhedron $X$ such that:

(1) $B$ is a locally finite covering of $X$,
(2) if $\beta, \gamma \in B$ then $\partial\beta, \beta \cap \gamma$ are unions of cells of $B$,
(3) if $\beta, \gamma$ are distinct cells of $B$, then $\operatorname{Int}\beta \cap \operatorname{Int}\gamma = \emptyset$.

We write $|B|$ for $X$ and do not distinguish between a cell $\beta$ of $B$ and the subcomplex it determines. A cell complex $B'$ is a **subdivision** of $B$ if $|B'| = |B|$ and every cell of $B$ is a union of cells of $B'$. A **based polyhedron** is a polyhedron with a preferred base-point; a **based cell complex** is a cell complex with a preferred vertex. All base-points will be denoted by 'bpt'.

Let $F$ be a polyhedron and let $B$ be a based cell complex. A **block bundle** $\xi$ over $B$ with **fibre** $F$ consists of a polyhedron $E(\xi)$ (the **total space** of $\xi$) with a closed sub-polyhedron $E_\beta(\xi)$ for each $\beta \in B$ and a $PL$ homeomorphism $b(\xi) : F \longrightarrow E_{\text{bpt}}(\xi)$, such that:

(1) $\{E_\beta(\xi) | \beta \in B\}$ is a locally finite covering of $E(\xi)$,
(2) if $\beta, \gamma \in B$ then

$$E_\beta(\xi) \cap E_\gamma(\xi) = \bigcup_{\delta \subset \beta\cap\gamma} E_\delta(\xi) ,$$

(3) if $\beta \in B$, there is a $PL$ homeomorphism $h : F \times \beta \longrightarrow E_\beta(\xi)$ such that

$$h(F \times \gamma) = E_\gamma(\xi) \quad (\gamma \subset \partial\beta) .$$

If $\xi$ is a block bundle over $B$ and $B_0$ is a subcomplex of $B$, the **restriction** $\xi|B_0$ is defined by

$$E(\xi|B_0) = \bigcup_{\beta \in B_0 \cup \{\text{bpt.}\}} E_\beta(\xi) ,$$

$$E_\beta(\xi|_{B_0}) = E_\beta(\xi) , \quad b(\xi|B_0) = b(\xi) .$$

Note that $\xi|B_0$ is a block bundle over $B_0 \cup \{\text{bpt.}\}$, not necessarily over $B_0$ itself.

If $\xi, \eta$ are block bundles over $B$, an **isomorphism** $h : \xi \longrightarrow \eta$ is a $PL$ homeomorphism $h : E(\xi) \longrightarrow E(\eta)$ such that

$$hE_\beta(\xi) = E_\beta(\eta) \quad (\beta \in B) , \quad hb(\xi) = b(\eta) .$$

A particular block bundle $\epsilon$ over $B$ is obtained by setting

$$E(\epsilon) = F \times B , \; E_\beta(\epsilon) = F \times \beta ,$$

$$b(\epsilon) = 1 \times \text{bpt} : F \longrightarrow F \times \text{bpt} .$$

A trivial block bundle is one isomorphic to $\epsilon$; an isomorphism $h : \epsilon \longrightarrow \xi$ is a **trivialisation** of $\xi$. It follows from condition (3) that $\xi|\beta$ is trivial for each $\beta \in B$.

Let $B, C$ be based cell complexes and let $\xi$ be a block bundle over $B$. Define a block bundle $\xi \times C$ over $B \times C$ by

$$E(\xi \times C) = E(\xi) \times C \;,\; E_{\beta\times\gamma}(\xi \times C) = E_\beta(\xi) \times \gamma$$

(for cells $\beta \in B, \gamma \in C$) and $b(\xi \times C) = b(\xi) \times \text{bpt}$.

**Lemma 1.** *Suppose $|B| = \beta$, where $\beta$ is an $n$-cell of $\beta$, and let $\gamma$ be an $(n-1)$-cell over $B$. If $\xi$ and $\eta$ are block bundles over $B$, any isomorphism $h : \xi|(\partial\beta - \gamma) \longrightarrow \eta|(\partial\beta - \gamma)$ can be extended to an isomorphism $h : \xi \longrightarrow \eta$.*

**Proof.** Since $\xi = \xi|\beta, \eta = \eta|\beta$, $\xi$ and $\eta$ are both trivial. Let $k$ and $l$ be trivialisations of $\xi, \eta$, respectively. Then a $PL$ homeomorphism

$$l^{-1}hk \;:\; F \times (\partial\beta - \gamma) \longrightarrow F \times (\partial\beta - \gamma)$$

is defined.

Choose a $PL$ homeomorphism

$$f : \; (\partial\beta - \gamma) \times I \longrightarrow B$$

such that $f_0 : (\partial\beta - \gamma) \longrightarrow B$ is the inclusion, and let

$$g \;=\; 1 \times f \;:\; F \times (\partial\beta - \gamma) \times I \longrightarrow F \times \beta.$$

The required extension of $f$ is given by

$$h \;=\; lg(l^{-1}hk \times I)g^{-1}k^{-1} \;:\; E(\xi) \longrightarrow E(\eta)\;.$$

**Lemma 2.** *Let $B$ be a based cell complex and take* bpt $\times\, 0$ *as base-point for $B \times I$. If $\xi, \eta$ are block bundles over $B \times I$, then any isomorphism*

$$h \;:\; \eta|(B \times 0) \cup (\text{bpt} \times I) \longrightarrow \eta|(B \times 0) \cup (\text{bpt} \times I)$$

*can be extended to an isomorphism $h : \xi \longrightarrow \eta$.*

**Proof.** Write $B^r$ for the $r$-skeleton of $B$, and let $C^r = (B \times 0) \cup (B^r \times I)$. Suppose inductively that $h$ can be extended to an isomorphism $h : \xi|C^r \longrightarrow \eta|C^r$; the induction starts trivially with $r = 0$. Let $\beta$ be an $(r+1)$-cell of $B$. By Lemma 1,

$$h \;:\; \eta|(\beta \times 0) \cup (\partial\beta \times I) \longrightarrow \eta|(\beta \times 0) \cup (\partial\beta \times I)$$

can be extended to an isomorphism $h : \xi|(\beta \times I) \longrightarrow \eta|(\beta \times I)$. Thus we have defined an isomorphism

$$h \;:\; \xi|C^r \cup (\beta \times I) \longrightarrow \eta|C^r \cup (\beta \times I)\;.$$

Do this for all $r$-cells of $B$ to obtain

$$h \;:\; \xi|C^{r+1} \longrightarrow \eta|C^{r+1}$$

extending the given isomorphism. The Lemma now follows by induction.

Let $\xi$ be a block bundle over $B$ and let $B'$ be a subdivision of $B$. A block bundle $\xi'$ over $B'$ is a subdivision of $\xi$ if $E(\xi') = E(\xi), E_{\beta'}(\xi') \subset E_\beta(\xi)$ (for all cells $\beta' \in B', \beta \in B$ with $\beta' \subset \beta$) and $b(\xi') = b(\xi)$.

**Theorem 1.** *Let $B'$ be a subdivision of a cell complex $B$. Any block bundle over $B'$ is a subdivision of some block bundle over $B$. Any block bundle $\xi$ over $B$ has a subdivision over $B'$, and any two subdivisions of $\xi$ over $B'$ are isomorphic.*

**Proof.** First we prove the following propositions together by induction on $n$.

$P_n$ : If $|B|$ is homeomorphic to an $n$-cell, then any block bundle over $B$ is trivial.

$Q_n$ : Let $\dim B \leq n$ and let $B_0$ be a subcomplex of $B$. Let $B'$ be a subdivision of $B$, inducing subdivision $B'_0$ of $B_0$. Let $\xi$ be a block bundle over $B$ and let $\xi'_0$ be a subdivision of $\xi|B_0$ over $B'_0$. Then there is a subdivision $\xi'$ of $\xi$ over $B'$ such that $\xi'_0 = \xi'|B'_0$.

Observe that $P_0$ and $Q_0$ are both true. We shall prove that $Q_n \Longrightarrow P_n$ and $P_n \,\&\, Q_n \Longrightarrow Q_{n+1}$.

**Proof that** $Q_n \Longrightarrow P_n$. Suppose $|B|$ is homeomorphic to an $n$-cell, and let $\xi$ be a block bundle over $B$. Since $|B|$ is collapsible, there is a simplicial subdivision $B'$ of $B$ which collapses simplicially to the base point [19]. Assuming $Q_n$, there is a subdivision $\xi'$ of $\xi$ over $B'$. It is enough to prove that $\xi'$ is trivial.

Let

$$B' \;=\; K_k \searrow^s K_{k-1} \searrow^s \dots \searrow^s K_0 \;=\; \{\text{bpt.}\}$$

be a sequence of elementary simplicial collapses. Suppose inductively that $\xi'|K_r$ is trivial; the induction starts with $r = 0$. Write

$$K_{r+1} \;=\; K_r \cup \Delta \;,\; K_r \cap \Delta \;=\; \Lambda \;,$$

where $\Delta$ is a simplex of $K_{r+1}$ and $\Lambda$ is the complement of a principal simplex in $\partial\Delta$. Let $h : F \times K_r \longrightarrow E(\xi'|K_r)$ be a trivialisation of $\xi'|K_r$. By Lemma 1, $h|F \times \Lambda$ extends to a trivialisation of $\xi|\Delta$. Thus we obtain a trivialisation of $\xi''|K_{r+1}$. By induction, $\xi'$ is trivial, as required.

**Proof that** $P_n \,\&\, Q_n \Longrightarrow Q_{n+1}$. Suppose $B, B_0, B', \xi, \xi'_0$ satisfy the hypotheses of $Q_{n+1}$. If $A$ is any subcomplex of $B$, we write $A'$ for the subdivision of $A$ induced by $B'$. Let $B_1 = B_0 \cup B^n$, assuming $Q_n$ there is a subdivision $\xi'_1$ of $\xi|B_1$ over $B'_1$ such that $\xi'_0|(B_0 \cap B^n)' = \xi'_1|(B_0 \cap B^n)'$. Let $\beta$ be an $(n+1)$-cell of $B - B_0$, and let

$\gamma$ be an $n$-cell of $B$ contained in $\partial\beta$. Since $|\partial\beta - \gamma|$ is homeomorphic to an $n$-cell, $\xi_1'|(\partial\beta - \gamma)'$ is trivial by $P_n$. Let $h$ be a trivialisation of $\xi_1'|(\partial\beta - \gamma)'$; a fortiori, $h$ is a trivialisation of $\xi|(\partial\beta - \gamma)$.

By Lemma 1, $h$ extends to a trivialisation of $\xi|\beta$. Let $C$ be the cell complex consisting of $\beta, \gamma$ and the cells of $(\partial\beta - \gamma)'$. Define a block bundle $\eta$ over $C$ by

$$E_\beta(\eta) = E_\beta(\xi) \ , \ E_\gamma(\eta) = E_\gamma(\xi)$$

and $E_{\delta'}(\eta) = E_{\delta'}(\xi_1')$ for each cell $\delta'$ of $(\partial\beta - \gamma)'$. Then $k$ is a trivialisation of $\eta$, so $\eta$ satisfies condition (3) in the definition of block bundle.

Let $\delta'$ be an $n$-cell of $(\partial\beta - \gamma)'$, so $|\partial\beta' - \delta'|$ is homeomorphic to an $n$-cell. Assuming $P_n, \xi_1'|(\partial\beta' - \delta')$ is trivial; let $h'$ be a trivialisation. A fortiori, $h'$ is a trivialisation of $\eta|(\partial\beta' - \delta')$.

By Lemma 1, $h'$ extends to a trivialisation $k'$ of $\eta$. In fact, $k'|F \times \partial\beta'$ is a trivialisation of $\xi_1'|\partial\beta'$, because $k'$ extends $h'$ and $k'(F \times \delta') = E_{\delta'}(\xi_1')$. To extend $\xi_1'|\partial\beta'$ to a subdivision $\xi'|\beta'$ of $\xi|\beta$, we define $E_{\alpha'}(\xi') = k'(F \times \alpha')$ for each cell $\alpha'$ of $\beta'$.

Do this for all $(n+1)$-cells of $B - B_0$, and define $\xi'|\beta = \xi_0'|\beta$ for each $(n+1)$-cell $\beta$ of $B_0$. We obtain a subdivision $\xi'$ of $\xi$ over $B'$ such that $\xi_0' = \xi'|B_0'$, as required.

By induction, $P_n$ and $Q_n$ are true for all $n$. Let $B$ be any based cell complex and let $B'$ be a subdivision of $B$.

Let $\xi'$ be a block bundle over $B'$. We define a block bundle $\xi$ over $B$ with $E(\xi) = E(\xi')$ by setting $E_\beta(\xi) = E(\xi'|\beta')$ (where $\beta'$ is the subdivision of $\beta$ induced by $B'$) for each cell $\beta$ of $B$. This clearly satisfies conditions (1),(2) in the definition of block bundle. By $P_n$, $\xi'|\beta'$ is trivial, so $\xi$ also satisfies condition (3). Clearly, $\xi'$ is a subdivision of $\xi$.

If $\xi$ is a block bundle over $B$, it follows from $Q_n$ (by induction on the skeleton of $B$) that $\xi$ has a subdivision over $B'$. Let $\xi_0', \xi_1'$ be two such subdivisions. Recall that $\eta = \xi \times I$ is a block bundle over $B \times I$. Define a block bundle $\eta_0'$ over $B' \times \partial I$ by $\xi_t' = \eta_0'|B' \times \{t\}$, $(t = 0, 1)$. Again it follows from $Q_n$ that $\eta$ has a subdivision $\eta'$ over $B' \times I$ such that $\eta_0' = \eta'|B' \times \partial I$. Observe that $\eta'|\text{bpt} \times I = \xi_0' \times I|\text{bpt} \times I$. By Lemma 2, the identity isomorphism

$$\eta'|(B \times 0) \cup (\text{bpt} \times I) \longrightarrow \xi_0' \times I|(B \times 0) \cup (\text{bpt} \times I)$$

extends to an isomorphism $\eta' \longrightarrow \xi_0' \times I$; it follows that $\xi_1' \cong \xi_0'$. This completes the proof of Theorem 1.

Let $X$ be a polyhedron and let $B, C$ be cell complexes with $|B| = |C| = X$; suppose all three have the same base-point. Let $\xi, \eta$ be block bundles over $B, C$ respectively. We call $\xi, \eta$ **equivalent** if, for some common subdivision $D$ of $B, C$, the subdivision $\xi$ over $D$ is isomorphic to the subdivision of $\eta$ over $D$. This relation is clearly reflexive and symmetric; by Theorem 1 it is also transitive.

Let $I_F(X)$ be the set of equivalence classes of block bundles over cell complexes $B$ with $|B| = X$. It is easily checked that, if $|B| = X$, then each member of $I_F(X)$ is represented by a unique isomorphism class of block bundles over $B$.

Suppose $X, Y$ are polyhedra and let $y \in I_F(Y)$. Let $B, C$ be cell complexes with $|B| = X, |C| = Y$, and let $\eta$ be a block bundle over $C$ representing $y$. If $p_2 : X \times Y \longrightarrow Y$ is the projection, let $p_2^*(y) \in I_F(X \times Y)$ be the equivalence class of $B \times \eta$.

If $i : X \longrightarrow Y$ is a closed based $PL$ embedding, let $C'$ be a subdivision of $C$ with a subcomplex $D'$ such that $|D'| = i(X)$. Let $\eta'$ be a subdivision of $\eta$ over $C'$, and let $\xi' = \eta'|D'$. It follows from Theorem 1 that the equivalence class $x' \in I_F(i(X))$ of $\xi'$ depends only on $y$. Let $i^*(y) \in I_F(X)$ correspond to $x'$ via the $PL$ homeomorphism $i : X \longrightarrow i(X)$. The next lemma will enable us to define $f^* : I_F(Y) \longrightarrow I_F(X)$ for any based $PL$ map $f : X \longrightarrow Y$.

**Lemma 3.** *Let $X, Y, V, W$ be polyhedra and let $i : X \longrightarrow V \times Y$, $j : X \longrightarrow W \times Y$ be closed based $PL$ embeddings such that $p_2 i = p_2 j : X \longrightarrow Y$. Then*

$$i^* p_2^* = j^* p_2^* : I_F(Y) \longrightarrow I_F(X) .$$

**Proof.** Let $k : X \longrightarrow V \times W \times Y$ be defined by

$$p_{13} k = i : X \longrightarrow V \times Y ,$$
$$p_{23} k = j : X \longrightarrow W \times Y .$$

In the diagram

$$\begin{array}{ccccc} & & I_F(V \times Y) & & \\ & \swarrow^{i^*} & \downarrow p_{13}^* & \nwarrow^{p_2^*} & \\ (1)\quad I_F(X) & \xleftarrow{k^*} & I_F(V \times W \times Y) & \xleftarrow{p_3^*} & I_F(Y) \\ & \nwarrow_{j^*} & \uparrow p_{23}^* & \swarrow_{p_2^*} & \\ & & I_F(W \times Y) & & \end{array}$$

the right-hand triangles are clearly commutative. We prove that the bottom left-hand triangle commutes. There is a contractible polyhedron $Z$ and a closed based $PL$ embedding $l : V \longrightarrow Z$. Consider the diagram

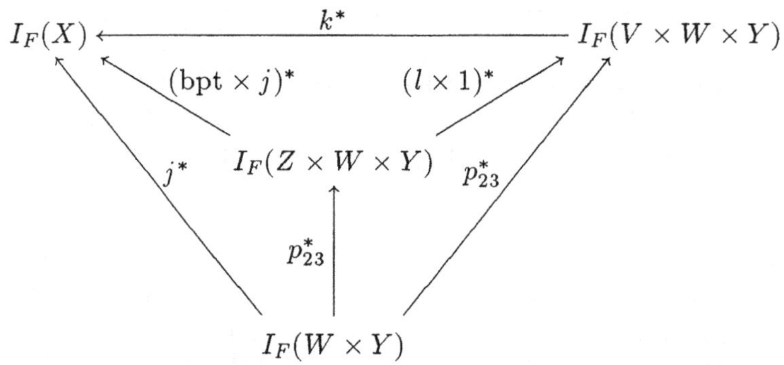


The bottom two triangles clearly commute. Since $Z$ is contractible to its base-point, $p_1(l \times 1)k \simeq p_1(\text{bpt} \times j)$. But $p_{23}(l \times 1)k = j = p_{23}(\text{bpt} \times j)$, so there is a closed based $PL$ isotopy between $(l \times 1)k$ and $(\text{bpt} \times j)$. It follows from Lemma 2 that $((l \times 1)k)^* = (\text{bpt} \times j)^*$. Clearly $k^*(l \times 1)^* = ((l \times 1)k)^*$, so the top triangle commutes. Therefore the bottom left-hand triangle in diagram (1) commutes, so the Lemma is proved.

Let $X$ and $Y$ be based polyhedra and let $f : X \longrightarrow Y$ be a based $PL$ map. There is a polyhedron $V$ and a factorization $f = p_2 i$, where $i : X \longrightarrow V \times Y$ is a closed based $PL$ embedding. For example, we can take $V = X$ and $i = 1 \times f$. By Lemma 3, the map $i^* p_2^* : I_F(Y) \longrightarrow I_F(X)$ depends only on $f$; we define $f^* = i^* p_2^*$.

**Lemma 4.** *$I_F$ is a contravariant functor from the category of based polyhedra and based $PL$ maps to the category of based sets.*

**Proof.** The base-point of $I_F(X)$ is the class of the trivial bundle. For any polyhedron $X$, $1_X^*$ is the identity map. Let $X, Y, Z$ be polyhedra, and let $f : X \longrightarrow Y$, $g : Y \longrightarrow Z$ be based $PL$ maps. Let $V, W$ be polyhedra and let $i : X \longrightarrow V \times Y$, $j : Y \longrightarrow W \times Z$ be closed based $PL$ embeddings such that $f = p_2 i, g = p_2 j$. Consider the diagram

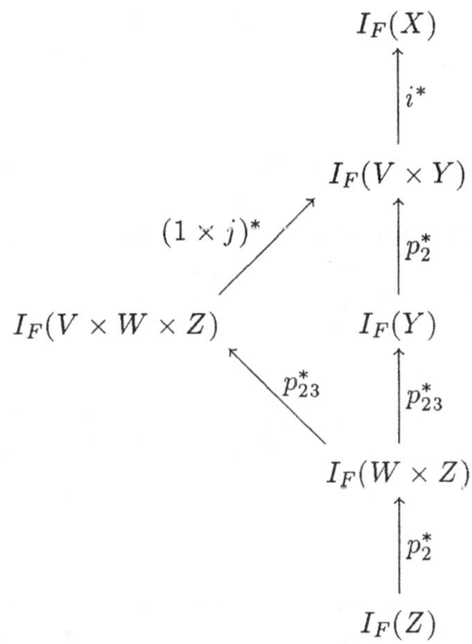

This clearly commutes; the right route defines $f^*g^*$ and the left route defines $(gf)^*$. This proves that $I_F$ is a contravariant functor.

**Theorem 2.** *If $F$ is compact, then there is a based polyhedron $\widetilde{BPL_F}$ and an element $w_I \in I_F(\widetilde{BPL_F})$ such that $f \mapsto f^*(w_I)$ defines a natural equivalence $[\ \ ,\widetilde{BPL_F}] \longrightarrow I_F$.*

**Proof.** First we show that $I_F$ satisfies the following axioms:

(1) If $X, Y$ are based polyhedra and $f_0 \simeq f_1 : X \longrightarrow Y$ by a based $PL$ homotopy, then $f_0^* = f_1^* : I_F(Y) \longrightarrow I_F(X)$.

(2) If $X_i$ is a based polyhedron $(i \in I)$ and $u_j : X_j \longrightarrow \bigvee_{i\in I} X_i$ is the inclusion, then $\Pi_{i\in I} u_i^* : I_F(\bigvee_{i\in I} X_i) \longrightarrow \Pi_{i\in I} I_F(X_i)$ is an isomorphism.

(3) Suppose that $X, X_0, X_1, X_2$ are polyhedra with $X = X_1 \cup X_2, X_0 = X_1 \cap X_2$, and that the inclusions $u_i : X_0 \longrightarrow X_i$, $v_i : X_i \longrightarrow X$ are based maps. If $x_i \in I_F(X_i)$, $(i = 1,\ 2)$ satisfy $u_1^*(x_1) = u_2^*(x_2)$, then there exists $x \in I_F(X)$ with $x_i = v_i^*(x)$, $(i = 1,\ 2)$.

(4) $I_F(S^0)$ is a single point and $I_F(S^n)$ is countable where $S^n$ denotes the boundary of an $(n+1)$-cell.

**Proof of (1).** This follows from Lemma 2 and a short argument about base-points.

**Proof of (2).** Let $B_i$ be a cell complex with $|B_i| = X_i$. Let $x \in \Pi_{i\in I} X_i$ and let $\xi_j$ be a block bundle over $B_j$ representing $p_j(x)$. Let $A = \cup_{i\in I} E(\xi_i)$, $A_0 = \cup_{i\in I} E_{\mathrm{bpt}}(\xi_i)$, $b = \cup_{i\in I} b(\xi)^{-1} : A_0 \longrightarrow F$ and define $E(\eta) = A \cup_b F$. If $\beta$ is a cell of

$\bigvee_{i \in I} B_i$, then $\beta$ is a cell of some $B_j$, so we can define $E_\beta(\eta) = E_\beta(\xi_j) \subset E(\eta)$. Let $b(\eta) = b(\xi_j) : F \longrightarrow E_{\mathrm{bpt}}(\eta)$, which is independent of $j$. Then $\eta$ is a block bundle over $\bigvee_{i \in I} B_i$; let $y \in I_F(\bigvee_{i \in I} X_i)$ be the class of $\eta$. Then $x \mapsto y$ defines an inverse to $\Pi_{i \in I} u_i^*$, so (2) is proved.

**Proof of (3).** Let $B$ be a cell complex with $|B| = X$ and with subcomplexes $B_0, B_1, B_2$ such that $|B_i| = X_i$ $(i = 0, 1, 2)$. Let $\xi_i$ be a block bundle over $B_i$ representing $x_i$ $(i = 1, 2)$. Since $u_1^*(x_1) = u_2^*(x_2)$, there is an isomorphism $h : \xi_1 | B_0 \longrightarrow \xi_2 | B_0$. Let $E(\xi) = E(\xi_1) \cup_h E(\xi_2)$, let $E_\beta(\xi) = E_\beta(\xi_i)$ if $\beta \in B_i$ and let $b(\xi) = b(\xi_1) = b(\xi_2) : F \longrightarrow E_{\mathrm{bpt}}(\xi)$. Then the class $x$ of $\xi$ has the required properties.

**Proof of (4).** Clearly $I_F(S^0)$ is a single point. Let $B$ be a cell complex such that $S^n = |B|$, and let $\beta$ be an $n$-cell of $B$. Any element $x \in I_F(S^n)$ can be represented by a block bundle $\xi$ over $B$. Let $k, l$ be trivialisations of $\xi | \beta$, $\xi | B - \beta$, and let $h = k^{-1} l : F \times \partial\beta \longrightarrow F \times \partial\beta$.

Since $F$ is compact, there are finite simplicial complexes $K, L$ with $|K| = F \times \beta, |L| = F \times (B - \beta)$ and such that $h$ is simplicial. Clearly the simplicial isomorphism class of the triple $(K,\ L,\ h)$ determines $x$ completely. But there are only countably many such classes (of triples), so $I_F(S^n)$ is countable.

Now we can apply Brown's Theorem on representable functors [4] to $I_F$. We deduce that there is a countable based $CW$ complex $W$ and a natural equivalence $R : [\ \ , W] \longrightarrow I_F$. By a theorem of J. H. C. Whitehead [18], there is a polyhedron $B\widetilde{PL}_F$ and a homotopy equivalence $\phi : B\widetilde{PL}_F \longrightarrow W$. Let $w_I = R(\phi) \in I_F(B\widetilde{PL}_F)$; then the pair $(B\widetilde{PL}_F,\ w_I)$ has the required properties.

**Remark.** The compactness of $F$ was only required to make the classifying space $B\widetilde{PL}_F$ a polyhedron. If $F$ were an infinite discrete space (for example), then the space $W$ constructed above would have uncountable fundamental group.

Our main concern is with block bundles having a compact $PL$ manifold $F$ as fibre. If $\xi$ is such a bundle over a cell complex $B$, we can define a block bundle $\partial\xi$ over $B$ with fibre $\partial F$ as follows.

Let $\beta$ be a cell of $B$, let $k, l$ be trivialisations of $\xi | B$ and let $h = k^{-1} l : F \times \beta \longrightarrow F \times \beta$. Since $h(F \times \gamma)$ for each $\gamma \subset \partial\beta$, $h(F \times \partial\beta) = F \times \partial\beta$. Therefore

$$h(\partial F \times \beta) \;=\; h(\overline{\partial(F \times \beta) - F \times \partial\beta}) \;=\; \partial F \times \beta\ ;$$

it follows that $k(\partial F \times \beta) = l(\partial F \times \beta)$. Define

$$E_\beta(\partial\xi) \;=\; k(\partial F \times \beta)\ ,$$

where $k$ is any trivialisation of $\xi|\beta$, and define

$$E(\partial\xi) = \bigcup_{\beta\in B} E_\beta(\partial\xi) \ , \ b(\partial\xi) = b(\xi)|\partial F \ .$$

Then $\partial\xi$ is a block bundle over $B$ with fibre $\partial F$.

**Lemma 5.** *Suppose that $|B|, F$ are compact PL manifolds, that $\partial\beta$ contains the base-point of $B$ and let $\xi$ bc a block bundle over $B$ with fibre $F$. Then $E(\xi)$ is a compact PL manifold and $\partial E(\xi) = E(\partial\xi) \cup E(\xi|\partial B)$.*

**Proof.** Let $B'$ be a simplicial division of $B$ and $\xi'$ be a subdivision of $\xi$ over $B'$. Clearly $\partial\xi'$ is then a subdivision of $\partial\xi$. If $p \in E(\xi)$, then

$$p \in P \ , \text{ where } P = E(\xi'|St(q,B')) - E(\xi'|Lk(q,B'))$$

for some vertex $q$ of $B'$; we can choose $q \in \text{Int}B$ unless $p \subset E(\xi|\partial B)$. Let

$$Q = St(q,B') - Lk(q,B') \ ,$$

so $Q$ is an open ball if $p \notin E(\xi|\partial B)$, and a half-open ball if $p \in E(\xi|\partial B)$.

A trivialisation $k$ of $\xi'|St(q,B')$ defines a homeomorphism $k : F \times Q \longrightarrow P$ such that $k(\partial F \times Q) = P \cap E(\partial\xi)$. Let $N$ be an open ball neighbourhood of $p_1 k^{-1}(p)$ in $F$ if $p \notin E(\partial\xi)$, or a half-open ball neighbourhood if $p \in E(\partial\xi)$.

If $p \notin E(\partial\xi) \cup E(\xi|\partial B)$, then $k(N \times Q)$ is an open ball neighbourhood of $p$ in $E(\xi)$. If $p \in E(\partial\xi) \cup E(\xi|\partial B)$, then $k(N \times Q)$ is a half-open ball neighbourhood of $p$, and $p \in k(\partial(N \times Q))$. This proves that $E(\xi)$ is a $PL$ manifold (obviously compact) with boundary $E(\partial\xi) \cup E(\xi|\partial B)$.

## II. Homotopy Properties of Block Bundles

Let $\xi$ be a block bundle over $B$ with fibre $F$. A **block fibration** for $\xi$ is a $PL$ map $\pi : E(\xi) \longrightarrow |B|$ such that $E_\beta(\xi) = \pi^{-1}(\beta)$ for each $\beta \in B$. A **block homotopy** for $\xi$ is a $PL$ map $H : E(\xi) \times I \longrightarrow |B|$ such that, for all $t \in I$, $H_t : E(\xi) \longrightarrow |B|$ is a block fibration for $\xi$.

**Lemma 6.** *Any block bundle $\xi$ has a block fibration, and any two block fibrations for $\xi$ are block homotopic.*

**Proof.** Write $B^r$ for the $r$-skeleton of $B$. There is a unique block fibration $\pi : E(\xi|B^0) \longrightarrow |B^0|$. Suppose inductively that $\pi$ can be extended to a block fibration $\pi : E(\xi|B^r) \longrightarrow |B^r|$, and let $\beta$ be an $(r+1)$-cell of $B$. Then $\pi : E(\xi|\partial\beta) \longrightarrow |\partial\beta|$ can be extended to a $PL$ map $\pi : E(\xi|\beta) \longrightarrow |\beta|$ such that $\pi^{-1}(|\partial\beta|) = E(\xi|\partial\beta)$. Do this for all $(r+1)$-cells of $B$ to obtain a block fibration $\pi : E(\xi|B^{r+1}) \longrightarrow |B^{r+1}|$

extending the given block fibration. By induction, $\xi$ has a block fibration; it is obvious that any two block fibrations for $\xi$ are block homotopic.

Let $\xi$ be a block bundle over $B$ with fibre $F$, and let $\pi$ be a block fibration for $\xi$. Let

$$\mathcal{E} \;=\; \{(x,\psi) : x \in E, \psi : I \longrightarrow |B| \text{ such that } \pi(x) = \psi(0)\}\ ,$$

with the compact open topology. Define $i : E(\xi) \longrightarrow \mathcal{E}$, $p : \mathcal{E} \longrightarrow |B|$ by $i(x) = (x, \text{constant})$ and $p(x,\psi) = \psi(1)$. Then $i$ is a homotopy equivalence and $p$ is a Hurewicz fibre map. Let $\mathcal{F} = p^{-1}(\text{bpt})$ be the fibre of $p$.

**Theorem 3.** *The map $ib(\xi) : F \longrightarrow \mathcal{F}$ is a homotopy equivalence.*

**Proof.** By [9], $\mathcal{F}$ has the homotopy type of a $CW$ complex. Choose a component $\mathcal{F}_1$ of $\mathcal{F}$; $\mathcal{F}_1$ lies in some component $\mathcal{E}_0$ of $\mathcal{E}$. Let $E_0$ be the corresponding component of $E(\xi)$, and let $B_0$ be the component of $B$ containing the base-point. It is easy to see that $\pi|E_0 \longrightarrow B_0$ must be surjective, so $F_0 = E_0 \cap E_{\text{bpt}}(\xi)$ is non-empty. Choose a base-point for $\mathcal{F}_0 = \mathcal{E}_0 \cap \mathcal{F}$.

If $n \geq 1$, there is a commutative diagram

$$\begin{array}{ccc} \pi_n(E_0, F_0) & \xrightarrow{\quad i_* \quad} & \pi_n(\mathcal{E}_0, \mathcal{F}_0) \\ & {\scriptstyle \pi_*}\searrow \qquad \swarrow{\scriptstyle p_*} & \\ & \pi_n(|B_0|, \text{bpt}) & \end{array}$$

Since $p : \mathcal{E}_0 \longrightarrow |B_0|$ is a Hurewicz fibration, $p_*$ is an isomorphism. Using the fact that $\pi : E_0 \longrightarrow |B_0|$ is a block fibration, we shall prove that $\pi_*$ is an isomorphism. It will follow that $i_*$ is an isomorphism; hence there is a unique component $F_1$ of $F_0$ with $i(F_1) \subset \mathcal{F}_1$. An application of the Five Lemma will show that $i_* : \pi_r(F_1) \longrightarrow \pi_r(\mathcal{F}_1)$ is an isomorphism for all $r \geq 1$, and the Theorem will follow by the Whitehead theorem.

To prove that $\pi_*$ is surjective, consider an element $\alpha \in \pi_n(|B_0|, \text{bpt})$. By subdividing, we may assume that $B_0$ is a simplicial complex (note that subdivision does not alter the homotopy class of $\pi : E_0, F_0 \longrightarrow |B_0|, \text{bpt}$). Let $D^n$ be a standard $n$-cell. There is a triangulation of $D^n$ such that $D^n \searrow^s \text{bpt} \in S^{n-1}$ and $\alpha$ is represented by a simplicial map $f : D^n, S^{n-1} \longrightarrow B_0, \text{bpt}$. Let

$$D^n \;=\; K_k \searrow^s K_{k-1} \searrow^s \ldots \searrow^s K_0 \;=\; \text{bpt}$$

be a sequence of elementary simplicial collapses.

Suppose inductively that there is a map $g : K_r \longrightarrow E_0$ such that, for all $x \in |K_r|$, $\pi g(x)$ is in the closed carrier of $f(x)$ in $B_0$. We can write $K_{r+1} = K_r \cup \Delta$, $K_r \cap \Delta = \Lambda$ for some simplex $\Delta \in K_{r+1}$. Let $\Delta_1 = \overline{\Delta - \Lambda}$, so $\Delta_1$ is a principal simplex of $\partial\Delta$. Let $\beta = f(\Delta)$, $\beta_1 = f(\Delta_1)$ be the image simplices in $B_0$. Then

$$g \ : \ \Lambda, \partial\Lambda \longrightarrow E_\beta(\xi), E_{\beta_1}(\xi)$$

is defined. Since $E_{\beta_1}(\xi)$ is a deformation retract of $E_\beta(\xi)$, $g$ can be extended to a map

$$g \ : \ \Delta, \Delta_1 \longrightarrow E_\beta(\xi), E_{\beta_1}(\xi) \ .$$

Thus we obtain an extension of $g$ to $g : K_{r+1} \longrightarrow E_0$ such that, for all $x \in |K_{r+1}|$, $\pi g(x)$ is in the closed carrier of $f(x)$ in $B_0$.

Now we have completed our induction and have obtained a map $g : D^n \longrightarrow E_0$ such that for all $x \subset D^n$, $\pi g(x)$ is in the closed carrier of $f(x)$ in $B_0$. In particular, $g(S^{n-1}) \subset F_0$, so $g$ represents an element $\beta \in \pi_n(E_0, F_0)$. Clearly $\pi_*\beta = \alpha$, so $\pi_*$ is injective as asserted. A similar argument shows that $\pi_*$ is injective, and the Theorem is proved.

We now restrict $F$ to be a compact $PL$ manifold with boundary $\partial F$. Let $X$ be a based polyhedron; a **Hurewicz fibration** over $X$ with fibre $(F, \partial F)$ consists of a pair of topological spaces $(\mathcal{E}, \partial\mathcal{E})$, a map $p : \mathcal{E} \longrightarrow X$ and a homotopy equivalence of pairs

$$b : F, \partial F \longrightarrow p^{-1}(\mathrm{bpt}), p^{-1}(\mathrm{bpt}) \cap \partial\mathcal{E}$$

such that;

(1) For all $x \in X$, $(p^{-1}(x), p^{-1}(x) \cap \partial\mathcal{E}) \simeq (F, \partial F)$,

(2) Given a pair of topological spaces $A, \partial A$, a map $f : A, \partial A \longrightarrow \mathcal{E}, \partial\mathcal{E}$ and a homotopy $G : A \times I \longrightarrow X$ such that $G_0 = pf$, then there exists a homotopy $H : A \times I, \partial A \times I \longrightarrow \mathcal{E}, \partial\mathcal{E}$ with $H_0 = f$, $G = pH$.

Two Hurewicz fibrations $(\mathcal{E}, \partial\mathcal{E}, p, b)$, $(\mathcal{E}', \partial\mathcal{E}', p', b')$ are **fibre homotopy equivalent** if there are maps

$$h \ : \ \mathcal{E}, \partial\mathcal{E} \longrightarrow \mathcal{E}', \partial\mathcal{E}' \quad , \quad h' \ : \ \mathcal{E}', \partial\mathcal{E}' \longrightarrow \mathcal{E}, \partial\mathcal{E}$$

and homotopies $H : h'h \simeq 1$, $H' : hh' \simeq 1$ such that, for all $t \in I$,

$$pH_t \ = \ p \ , \quad H_t b \ = \ b \ , \quad p'H'_t \ = \ p' \ , \quad H'_t b' \ = \ b' \ .$$

We write $H_F(X)$ for the set of fibre homotopy equivalence classes of Hurewicz fibrations over $X$ with fibre $(F, \partial F)$. The well-known construction for induced fibrations makes $H_F$ into a contravariant functor from the category of based polyhedra and based $PL$ maps to the category of based sets. A proof that $H_F$ is representable is indicated in [4]; the step which is given without proof can be dealt with by the methods of Theorem 3 above. We summarise the conclusion as follows.

**Proposition.** *If $F$ is a compact PL manifold, then there is a based polyhedron $BG_F$ and an element $w_H \in H_F(BG_F)$ such that $f \mapsto f^*(w_H)$ defines a natural equivalence $[\ ,BG_F] \longrightarrow H_F$.*

**Lemma 7.** *There is a natural transformation $S : I_F \longrightarrow H_F$.*

**Construction of $S$.** Let $X$ be a based polyhedron and let $x \in I_F(X)$. Let $B$ be a cell complex with $|B| = X$, and let $\xi$ be a block bundle over $B$ representing $x$. By Lemma 6, $\xi$ has a block fibration $\pi : E(\xi) \longrightarrow X$. Construct $p : \mathcal{E} \longrightarrow X$ as above, and let $\partial\mathcal{E} = \{(x, \psi) \in \mathcal{E} : x \in E(\partial\xi)\}$. It is easily proved that $p : \mathcal{E}, \partial\mathcal{E} \longrightarrow X$ satisfies part (2) of the definition of Hurewicz fibration. By Theorem 3, part (1) is also satisfied, and

$$ib(\xi) \ :\ F, \partial F \longrightarrow p^{-1}(\mathrm{bpt}), p^{-1}(\mathrm{bpt}) \cap \partial\mathcal{E}$$

is a homotopy equivalence. Therefore $(\mathcal{E}, \partial\mathcal{E}, p, ib(\xi))$ defines an element $S(\xi, \pi) \in H_F(X)$.

Let $\pi'$ be another block fibration for $\xi$. Construct $(\mathcal{E}', \partial\mathcal{E}', p', i')$ from $\pi'$ as above. Define $j' : \mathcal{E}', \partial\mathcal{E}' \longrightarrow E(\xi), E(\partial\xi)$ by $j'(x, \psi) = x$; then $j'$ is a homotopy inverse to $i'$. Thus $ij' : \mathcal{E}', \partial\mathcal{E}' \longrightarrow \mathcal{E}, \partial\mathcal{E}$ is a homotopy equivalence of pairs and $p' \simeq p.ij'$ via a homotopy $H$ with $H_t.i'b(\xi) = ib(\xi)$. It follows from Theorem 6.1 of [5] (modified to take account of base-points and pairs of fibres) that $(\mathcal{E}', \partial\mathcal{E}', p', i'b(\xi))$ is fibre homotopy equivalent to $(\mathcal{E}, \partial\mathcal{E}, p, ib(\xi))$. Therefore $S(\xi, \pi)$ depends only on $\xi$.

If $\xi'$ is a subdivision of $\xi$, then $S(\xi, \pi) = S(\xi, \pi') = S(\xi', \pi')$ for any block fibrations $\pi, \pi'$ of $\xi, \xi'$. Therefore $S(\xi, \pi)$ depends only on the equivalence class $x$ of $\xi$; we write $S(x) = S(\xi, \pi)$.

**Naturality of $S$.** It is enough to prove that $S$ is natural
(1) with respect to projections $p_2 : Y \times X \longrightarrow X$,
(2) with respect to closed based $PL$ embeddings $j : Y \longrightarrow X$.

**Proof of (1).** Let $B, C$ be cell complexes with $|B| = X$, $|C| = Y$. Let $\xi$ be a block bundle over $B$ representing $x \in I_F(X)$, and let $\pi$ be a block fibration for $\xi$. Then $\pi \times 1 : E(\xi) \times Y \longrightarrow X \times Y$ is a block fibration for $\xi \times C$ (which represents $p_2^*(x)$). Construct $(\mathcal{E}, \partial\mathcal{E}, p, ib(\xi))$ representing $S(x)$. Let $Y^I$ be the space of **unbased** maps $\psi : I \longrightarrow Y$ and define $e_1 : Y^I \longrightarrow Y$ by $e_1(\psi) = \psi(1)$. Then

$$(\mathcal{E} \times Y^I, \partial\mathcal{E} \times Y^I, p \times e_1, (i \times \mathrm{bpt})b(\xi))$$

represents $S(p_2^*(x))$. But this fibration is equivalent to

$$(\mathcal{E} \times Y, \partial\mathcal{E} \times Y, p \times 1, (i \times \mathrm{bpt})b(\xi)) ,$$

which represents $p_2^*(S(x))$.

**Proof of (2).** Let $B$ be a cell complex with $|B| = X$ and with a subcomplex $C$ such that $|C| = j(Y)$. Let $\xi$ be a block bundle over $B$ representing $x \in I_F(X)$, and let $\pi$ be a block fibration for $\xi$. then $\pi|E(\xi|C) \longrightarrow |C|$ is a block fibration for $\xi|C$ (which represents $j^*(x)$). Construct $(\mathcal{E}, \partial\mathcal{E}, p, ib(\xi))$ representing $S(x)$. Write $i'$ for the restriction

$$i| \; : \; E(\xi|C), E(\partial\xi|C) \longrightarrow p^{-1}|C|, p^{-1}|C| \cap \partial\mathcal{E} \; ;$$

clearly $\pi|E(\xi|C) = pi'$.

Identify $F$ with $b(\xi)F$ and write $\mathcal{F}$ for $p^{-1}(\text{bpt})$. Consider the commutative diagram

$$\begin{array}{ccc} \pi_n(E(\xi|C), F) & \xrightarrow{\quad i'_* \quad} & \pi_n(p^{-1}|C|, \mathfrak{F}) \\ & {\scriptstyle \pi_*} \searrow \quad \swarrow {\scriptstyle p_*} & \\ & \pi_n(|C|, \text{bpt}) & \end{array}$$

As in Theorem 3, $\pi_*$ is an isomorphism; since $p_*$ is an isomorphism, $i'_*$ is also an isomorphism. Therefore $i' : E(\xi|C) \longrightarrow p^{-1}|C|$ is a homotopy equivalence. Similarly $i' : E(\partial\xi|C) \longrightarrow p^{-1}|C| \cap \partial\mathcal{E}$ is a homotopy equivalence, so $i'$ is a homotopy equivalence of pairs. It follows from Theorem 6.1 of [5] that

$$(p^{-1}|C|, p^{-1}|C| \cap \partial\mathcal{E}, p|p^{-1}|C|, ib(\xi))$$

represents $S(j^*x)$, so $j^*S(x) = S(j^*x)$. This completes the proof of Lemma 7.

Recall that $w_I \in I_F(B\widetilde{PL_F})$, $w_H \in H_F(BG_F)$ are the universal elements. There is a based map $\chi : B\widetilde{PL}_F \longrightarrow BG_F$ such that $S(w_I) = \chi^*(w_H)$. This defines the based homotopy class of $\chi$ uniquely.

Consider the topological space $L = \{(x, \psi)\}$ of pairs with $x \in B\widetilde{PL}_F$, $\psi : I \longrightarrow BG_F$ such that $\chi(x) = \psi(0)$, $\psi(1) = \text{bpt}$, with (bpt,constant) as base-point. There is a based map $\chi' : L \longrightarrow B\widetilde{PL}_F$ defined by $\chi'(x, \psi) = x$. By theorems of Milnor [9] and J. H. C. Whitehead [18], there is a based polyhedron $G_F/\widetilde{PL}_F$ and a homotopy equivalence $i : G_F/\widetilde{PL}_F \longrightarrow L$. Define $\chi_1 = \chi' i : G_F/\widetilde{PL}_F \longrightarrow B\widetilde{PL}_F$.

Let $B$ be a based cell complex and let $F$ be a compact $PL$ manifold. A **$G_F/\widetilde{PL}_F$-bundle** over $B$ consists of a block bundle $\xi$ over $B$ with fibre $F$ and a $PL$ map

$$t \; : \; E(\xi), E(\partial\xi) \longrightarrow F, \partial F$$

such that $tb(\xi) = 1$. Two $G_F/\widetilde{PL}_F$-bundles $(\xi, t)$ and $(\eta, u)$ over $B$ are **isomorphic** if there is an isomorphism $h : \xi \longrightarrow \eta$ such that $uh \simeq t$ (rel $b(\xi)(F)$). Define the **equivalence** of $G_F/\widetilde{PL}_F$-bundles over a polyhedron $X$ as in Chapter I, and let $J_F(X)$ be the set of equivalence classes.

**Lemma 8.** *Let $(\xi, t)$ be a $G_F/\widetilde{PL}_F$-bundle over $B$ and let $\pi : E(\xi) \longrightarrow |B|$ be a block fibration for $\xi$. Then*

$$t \times \pi \ : \ E(\xi), E(\partial\xi) \longrightarrow F \times |B|, \partial F \times |B|$$

*is a homotopy equivalence of pairs.*

**Proof.** Apply Theorem 3 as in the proof of Lemma 7.

We make $J_F$ into a contravariant functor as follows. Let $f : X \longrightarrow Y$ be a based $PL$ map, and suppose $f = p_2 j$, where $j : X \longrightarrow V \times Y$ is a closed based $PL$ embedding. Let $B$, $C$, $D$ be cell complexes with

$$|B| = X \ , \ |C| = Y \ , \ |D| = Z \ ,$$

and let $(\eta, u)$ be a $G_F/\widetilde{PL}_F$-bundle over $C$ representing $y \in J_F(Y)$. Let $(D \times C)'$ be a subdivision of $D \times C$ with $j(B)$ as a subcomplex, and let $(D \times \eta)'$ be a subdivision of $D \times \eta$ over $(D \times C)'$. Then $f^*(y)$ is represented by

$$((C \times \eta)'|j(B) \ , \ up_2|E((C \times \eta)'|j(B))) \ .$$

The proof of Lemma 3 shows that $f^* : J_F(Y) \longrightarrow J_F(X)$ is well-defined, and that $J_F$ is a contravariant functor.

**Theorem 4.** *If $F$ is a compact $PL$ manifold, then there is an element $w_J \in J_F(G_F/\widetilde{PL}_F)$ such that $f \mapsto f^*(w_J)$ defines a natural equivalence $[\ , G_F/\widetilde{PL}_F] \longrightarrow J_F$.*

**Proof.** Let $C_I$, $C_J$, $C_H$ be cell complexes with

$$|C_I| = B\widetilde{PL}_F \ , \ |C_J| = G_F/\widetilde{PL}_F \text{ and } |C_H| = BG_F \ .$$

Let $\eta_I$ be a block bundle over $C_I$ representing $w_I$, and let $\eta_H$ be a Hurewicz fibration over $|C_H|$ representing $w_H$. Let $\eta_J$ be a block bundle over $C_J$ representing $\chi_1^*(w_I)$, and let $\pi_I$, $\pi_J$ be block fibrations for $\eta_I$, $\eta_J$.

Recall that $S(w_I) = \chi^*(w_H)$; let $h : S(\eta_I, \pi_I) \longrightarrow \chi^*(\eta_H)$ be a fibre homotopy equivalence. The proof of naturality of $S$ (Lemma 7) provides a fibre homotopy equivalence

$$S(\eta_J, \pi_J) \longrightarrow \chi_1^* S(\eta_I, \pi_I) \ .$$

Compose this with

$$\chi_1^* h : \chi_1^* S(\eta_I, \pi_I) \longrightarrow \chi_1^* \chi^*(\eta_H)$$

to obtain a fibre homotopy equivalence

$$h_1 \ : \ S(\eta_J, \pi_J) \longrightarrow \chi_1^* \chi^*(\eta_H) \ .$$

Now $\chi\chi_1 = \chi\chi' i$, where $\chi' : L \longrightarrow BG_F$ sends $(x, \psi)$ to $x$. There is an obvious null-homotopy $H : L \times I \longrightarrow BG_F$ of $\chi\chi'$, so $H' = H(i \times 1)$ is a null-homotopy of $\chi\chi_1$. Let $h' : \chi_1^* \chi^*(\eta_H) \longrightarrow \epsilon$ be the trivialisation defined by $H'$. The composite

$$E(\eta_J) \xrightarrow{i} S(\eta_J, \pi_J) \xrightarrow{h_1} \chi_1^* \chi^*(\eta_H) \xrightarrow{h'} \epsilon \xrightarrow{p_1} F$$

defines a map

$$u_J : E(\eta_J), E(\partial \eta_J) \longrightarrow F, \partial F$$

such that $u_J b(\eta_J) = 1$. Let $w_J$ be the equivalence class of $(\eta_J, u_J)$ in $J_F(G_F/\widetilde{PL}_F)$.

Clearly $f \mapsto f^*(w_J)$ defines a natural transformation from $[\ \ , G_F/\widetilde{PL}_F]$ to $J_F$. Let $B$ be a cell complex and let $(\xi, t)$ be a $G_F/\widetilde{PL}_F$-bundle over $B$; we have to prove that the equivalence class of $(\xi, t)$ corresponds to a unique element of $[|B|, G_F/\widetilde{PL}_F]$. Let $\pi$ be a block fibration for $\xi$.

There is a map $g : |B| \longrightarrow B\widetilde{PL}_F$ such that $\xi$ represents $g^*(w_I)$; $g$ is unique up to homotopy. The proof of naturality of $S$ (Lemma 7) provides a fibre homotopy equivalence $S(\xi, \pi) \longrightarrow g^* S(\eta_I, \pi_I)$. Compose this with $g^* h : g^* S(\eta_I, \pi_I) \longrightarrow g^* \chi^*(\eta_H)$ to obtain a fibre homotopy equivalence $k : S(\xi, \pi) \longrightarrow g^* \chi^*(\eta_H)$.

Now $tk^{-1} : g^* \chi^*(\eta_H) \longrightarrow F$ defines a fibre homotopy trivialisation of $g^* \chi^*(\eta_H)$, unique up to fibre homotopy. Let $K : |B| \times I \longrightarrow BG_F$ be the corresponding null-homotopy of $\chi g : |B| \longrightarrow BG_F$. Then $(g, K)$ defines the unique homotopy class of maps $f : |B| \longrightarrow G_F/\widetilde{PL}_F$ such that $(\eta, t)$ represents $f^*(w_J)$. This completes the proof of Theorem 4.

## III. Tangential Properties of Block Bundles

Let $I^n$ denote the product of $n$ copies of the unit interval; we write $G_n/\widetilde{PL}_n$ for $G_{I^n}/\widetilde{PL}_{I^n}$. The obvious natural transformation $J_{I^n} \longrightarrow J_{I^{n+1}}$ (multiply the fibre of each bundle by $I$) defines a homotopy class of maps $G_n/\widetilde{PL}_n \xrightarrow{i_n} G_{n+1}/\widetilde{PL}_{n+1}$. Write $G/PL$ for the direct limit of the sequence

$$\xrightarrow{i_{n-1}} G_n/\widetilde{PL}_n \xrightarrow{i_n} G_{n+1}/\widetilde{PL}_{n+1} \xrightarrow{i_{n+1}} \cdots .$$

More precisely, for $n = 1, 2, 3, \ldots$ replace $G_{n+1}/\widetilde{PL}_{n+1}$ by a homotopy equivalent polyhedron in such a way that $i_n$ is an injection, and identify $G_n/\widetilde{PL}_n$ with $i_n(G_n/\widetilde{PL}_n)$. Now define $G/PL$ to be the nested union of the $G_n/\widetilde{PL}_n$; it can be shown that the homotopy type of $G/PL$ is independent of the choices made (see Lemma 1.7 of [3]).

$G/PL$ was studied by Sullivan in his thesis (but he called it $F/PL$). The aim of this chapter is to obtain a map $\theta : G_F/\widetilde{PL}_F \longrightarrow (G/PL)^F$, where $(G/PL)^F$

is the space of all **unbased** maps from $F$ into $G/PL$ (with the compact open topology). Let $\mathcal{C}$ be the category of based, compact, stably parallelizable $PL$ manifolds and based $PL$ maps. Our first step is to define a natural transformation

$$T : [\ ,G_F/\widetilde{PL}_F] \longrightarrow [\ ,(G/PL)^F]\,,$$

where the functors are defined on $\mathcal{C}$.

Let $N$ be an object of $\mathcal{C}$, with boundary $\partial N$, and let $B$ be a cell complex with $|B| = N$. Let $\beta$ be a principal cell of $B$ with the base-point as one vertex. Let $x \in [N, G_F/\widetilde{PL}_F]$ be represented by a $G_F/\widetilde{PL}_F$-bundle $(\xi, t)$ over $B$. Extend $b(\xi)p_1 : F \times \mathrm{bpt} \longrightarrow E(\xi|\mathrm{bpt})$ to a homeomorphism $b : F \times \beta \longrightarrow E(\xi|\beta)$. Change $t$ by a homotopy (rel $b(\xi)(F)$) until $tb = p_1 : F \times \beta \longrightarrow F$.

We write $E = E(\xi)$, so $E$ is a $PL$ manifold with $\partial E = E(\partial\xi) \cup E(\xi|\partial B)$. We write $W$ for $F \times \beta$ and identify $W$ with $b(W)$. Let $\pi : E \longrightarrow N$ be a block fibration such that $\pi|F \times \beta = p_2$. Let $Q = F \times N$, so by Lemma 8, $t \times \pi : E, \partial E \longrightarrow Q, \partial Q$ is a homotopy equivalence of pairs. Note that $t \times \pi|W = 1$ and $(t \times \pi)^{-1}(W) = W$. Let $g : Q, \partial Q \longrightarrow E, \partial E$ be a homotopy inverse to $t \times \pi$ such that $g|W = 1$ and $g^{-1}(W) = W$.

Let $k$ be large, and choose embeddings

$$e : E, \partial E \longrightarrow D^k, S^{k-1} \quad , \quad q : Q, \partial Q \longrightarrow D^k, S^{k-1}$$

such that $e|W = q|W$. By [6], there exists normal bundles $\nu_Q$, $\nu_E$ of $Q$, $E$ in $D^k$. Choose $\nu_Q$, $\nu_E$ so that $\nu_Q|W = \nu_E|W$ (using the uniqueness theorem of [6] and regular neighbourhood theory). Let $Q^\nu$, $E^\nu$, $W^\nu$ be Thom spaces for $\nu_Q$, $\nu_E$, $\nu_Q|W$, and let

$$\gamma : Q^\nu/\partial Q^\nu \longrightarrow W^\nu/\partial W^\nu \quad , \quad \gamma : E^\nu/\partial E^\nu \longrightarrow W^\nu/\partial W^\nu$$

be the collapsing maps. Let $\bar{\nu}_Q = g^*(\nu_E)$ have Thom space $Q^{\bar{\nu}}$ and collapsing map

$$\bar{\gamma} : Q^{\bar{\nu}}/\partial Q^{\bar{\nu}} \longrightarrow W^{\bar{\nu}}/\partial W^{\bar{\nu}} = W^\nu/\partial W^\nu\,.$$

There is a homotopy equivalence $\bar{h} : E^\nu/\partial E^\nu \longrightarrow Q^{\bar{\nu}}/\partial Q^{\bar{\nu}}$ covering $t \times \pi : E \longrightarrow Q$ and such that $\bar{\gamma}\bar{h} = \gamma$.

There is a map $D^k \longrightarrow Q^\nu/\partial Q^\nu$ which collapses

$$S^{k-1} \cup (\text{complement of total space of } \nu_Q)$$

to a point. If we identify $S^k$ with $D^k/S^{k-1}$, we obtain a map $\phi : S^k \longrightarrow Q^\nu/\partial Q^\nu$; let $\psi : S^k \longrightarrow E^\nu/\partial E^\nu$ be defined similarly. Let $\bar{\phi} = \bar{h}\psi : S^k \longrightarrow Q^{\bar{\nu}}/\partial Q^{\bar{\nu}}$; then $\bar{\gamma}\bar{\phi} = \gamma\phi = \gamma\psi$.

By theorems of Atiyah [1] and Wall [15, Th 3.5] there is a fibre homotopy equivalence $\bar{f} : \bar{\nu}_Q \longrightarrow \nu_Q$ such that $\bar{f}\bar{\phi} \simeq \phi$. It follows from Wall's theorem that $\bar{f}$ is unique up to fibre homotopy. Consider $\tilde{f} = \bar{f}|(\bar{\nu}_Q|W) \longrightarrow (\nu_Q|W)$; this has the property that

$$\gamma\phi \simeq \gamma\bar{f}\bar{\phi} = \tilde{f}(\bar{\gamma}\bar{\phi}) = \tilde{f}(\gamma\phi) .$$

By the uniqueness clause in Wall's theorem, $\tilde{f}$ is fibre homotopic to the identity. Therefore we can alter $\bar{f}$ by a fibre homotopy until it is the identity on $\bar{\nu}_Q|W$.

Let $G$ be defined as in [8] (this agrees with the definition used in [15]), so $G$ is an $H$-space. Since $W$ is a retract of $Q$, the map $[Q/W,G]\longrightarrow[Q,G]$ is injective. It follows that two fibre equivalences $\bar{f}_0, \bar{f}_1 : \bar{\nu}_Q\longrightarrow\nu_Q$ which are the identity on $\bar{\nu}_Q|W$ are fibre homotopic (rel $\bar{\nu}_Q|W$) if and only if they are fibre homotopic. Therefore the fibre homotopy equivalence $\bar{f} : \bar{\nu}_Q\longrightarrow\nu_Q$ obtained above is unique up to fibre homotopy (rel $\bar{\nu}_Q|W$).

Let $\tau_Q$ be the tangent bundle on $Q$, and choose a fixed trivialisation $\kappa : \tau_Q\oplus\nu_Q\longrightarrow\epsilon$. Then

$$f = \kappa(1\oplus\bar{f}) : \tau_Q\oplus\bar{\nu}_Q \longrightarrow \epsilon$$

is a fibre homotopy equivalence, which agrees with $\kappa$ on $\tau_Q\oplus\bar{\nu}_Q|W$. The pair $(\tau_Q\oplus\bar{\nu}_Q, f)$ represents an element

$$T(x) \in [Q/W, G/PL] \cong [N, (G/PL)^F] .$$

Since the normal invariants $\phi$, $\psi$ are unique up to homotopy and $PL$ bundle automorphisms, $T(x)$ depends only on $x$. Thus we have defined a map

$$T : [N, G_F/\widetilde{PL}_F] \longrightarrow [N, (G/PL)^F] .$$

**Lemma 9.** *$T$ is a natural transformation (between functors from $C$ to the category of based sets).*

**Proof.** Let $f : M\longrightarrow N$ be a based $PL$ map. Express $f$ as a composite

$$M \xrightarrow{\times 0} M\times D^r \xrightarrow{u} N\times D^s \xrightarrow{p_1} N ,$$

where $u$ is a codimension 0 embedding. We prove that $T$ is natural

(1) with respect to $\times 0$ and $p_1$,
(2) with respect to codimension 0 embeddings.

**Proof of 1.** Consider $p_1 : N\times D^s\longrightarrow N$; let $B$ be a cell complex with $|B| = N$. Let $(\xi, t)$ be a $G_F/\widetilde{PL}_F$-bundle over $B$ representing $x\in[N, G_F/\widetilde{PL}_F]$, so that $(\xi\times D^s, tp_1)$ represents $p_1^*(x)$. Let

$$Q , W , \nu_Q , \bar{\nu}_Q , \phi : S^k \longrightarrow Q^\nu/\partial Q^\nu , \bar{\phi} : S^k \longrightarrow Q^{\bar{\nu}}/\partial Q^{\bar{\nu}}$$

be defined for $(\xi, t)$ as above. The corresponding objects for $(\xi \times D^s, tp_1)$ are

$$Q_s = Q \times D^s \ , \ W_s = W \times D^s \ ,$$
$$\nu_{Q_s} = \nu_Q \times D^s \ , \ \bar{\nu}_{Q_s} = \bar{\nu}_Q \times D^s \ ,$$
$$S^s\phi : S^{s+k} \longrightarrow Q_s^\nu/\partial Q_s^\nu \ , \ S^s\bar{\phi} : S^k \longrightarrow Q_s^{\bar{\nu}}/\partial Q_s^{\bar{\nu}}$$

(note that $Q_s^\nu/\partial Q_s^\nu \cong S^s(Q^\nu/\partial Q^\nu)$, $Q_s^{\bar{\nu}}/\partial Q_s^{\bar{\nu}} \cong S^s(Q^{\bar{\nu}}/\partial Q^{\bar{\nu}})$).

Let $\bar{f} : \bar{\nu}_Q \longrightarrow \nu_Q$ be a fibre homotopy equivalence such that $\bar{f}\bar{\phi} \simeq \phi$ and $\bar{f}$ is the identity on $\bar{\nu}_Q|W$. Then $\bar{f}_s = \bar{f} \times 1 : \bar{\nu}_{Q_s} \longrightarrow \nu_{Q_s}$ is the identity on $\bar{\nu}_{Q_s}|W$, and $\bar{f}_s(S^s\bar{\phi}) \simeq S^s\phi$. Therefore $(\tau_{Q_s} \oplus \bar{\nu}_{Q_s}, 1 \oplus \bar{f}_s)$ represents $T(p_1^*(x))$. It follows that $T(p_1^*(x)) = p_1^*(T(x))$, as required. Since $\times 0 : N \longrightarrow N \times D^s$ is a homotopy inverse to $p_1$, $T$ is also natural with respect to $\times 0$.

**Proof of 2.** Let $u : M \longrightarrow N$ be a codimension 0 embedding. Let $B$ be a cell complex with $|B| = N$ and with a subcomplex $A$ such that $|A| = u(M)$. Choose $\beta$ to be a cell of $A$ containing the base-point, as above. Let $(\xi, t)$ be a $G_F/\widetilde{PL}_F$-bundle over $B$ representing $x \in [N, G_F/\widetilde{PL}_F]$; then $(\xi|A, t|E(\xi|A))$ represents $u^*(x)$. Let

$$E = E(\xi) \ , \ D = D(\xi|A) \ , \ Q = F \times N \ , \ P = F \times M \ .$$

Identify $W$ with $b(W) \subset D \subset E$, as above. Let

$$g : Q, P, \partial Q, \partial P \longrightarrow E, D, \partial E, \partial D$$

be a homotopy inverse to $t \times \pi$ such that $g|W$ is the identity.

Choose embeddings

$$g : Q, \partial Q \longrightarrow D^k, S^{k-1} \ , \ e : E, \partial E \longrightarrow D^k, S^{k-1}$$

agreeing on $W$, as above. Let $\nu_Q$, $\nu_E$ be normal bundles with $\nu_Q|W = \nu_E|W$, and let $\nu_P = \nu_Q|P$, $\nu_D = \nu_E|D$. We obtain collapsing maps

$$\eta : Q^\nu/\partial Q^\nu \longrightarrow P^\nu/\partial P^\nu \ , \ \eta : E^\nu/\partial E^\nu \longrightarrow D^\nu/\partial D^\nu \ .$$

Let $\bar{\nu}_Q = g^*(\nu_E)$, let $\bar{\nu}_P = \bar{\nu}_Q|P$ and let $\bar{h} : E^\nu/\partial E^\nu \longrightarrow Q^{\bar{\nu}}/\partial Q^{\bar{\nu}}$ be a homotopy equivalence covering $t \times \pi : E \longrightarrow Q$, such that $\bar{\gamma}\bar{h} = \gamma$ (where $\gamma$, $\bar{\gamma}$ are as above).

If

$$\phi : S^k \longrightarrow Q^\nu/\partial Q^\nu \ , \ \psi : S^k \longrightarrow E^\nu/\partial E^\nu$$

are collapsing maps for $Q$, $E$, then $\eta\phi$, $\eta\psi$ are collapsing maps for $P$, $D$. Let $\bar{\phi} = \bar{h}\psi : S^k \longrightarrow Q^{\bar{\nu}}/\partial Q^{\bar{\nu}}$; the corresponding map for $P$ is $\bar{h}\eta\psi : S^k \longrightarrow P^{\bar{\nu}}/\partial P^{\bar{\nu}}$. Let $\bar{f} : \bar{\nu}_Q \longrightarrow \nu_Q$ be a fibre homotopy equivalence such that $\bar{f}$ is the identity on $\bar{\nu}_Q|W$ and $\bar{f}\bar{\phi} \simeq \phi$.

Now $\tilde{f} = \bar{f}|\bar{\nu}_P \longrightarrow \nu_P$ is a fibre homotopy such that

$$\tilde{f}(\bar{h}\eta\psi) = \tilde{f}(\eta\bar{\phi}) = \eta\bar{f}\bar{\phi} \simeq \eta\phi$$

and $\tilde{f}$ is the identity on $\bar{\nu}_P|W$. Therefore $T(x)$, $T(u^*(x))$ are represented by $(\tau_Q \oplus \bar{\nu}_Q, 1 \oplus \bar{f})$, $(\tau_P \oplus \bar{\nu}_P, 1 \oplus \tilde{f})$ respectively. It follows that $T(u^*(x)) = u^*(T(x))$, as required. This proves the Lemma.

Since $G_F/\widetilde{PL}_F$ and $(G/PL)^F$ have the homotopy type of countable $CW$ complexes, it follows from Lemma 1.7 of [3] that there is a map $\theta : G_F/\widetilde{PL}_F \longrightarrow (G/PL)^F$ such that $T = \theta_*$. Unfortunately, the homotopy class of $\theta$ is not uniquely determined by this condition.

**Theorem 5.** *Let $F^n$ be a closed 1-connected $PL$ manifold with $n \geq 4$. Let $F^* = \overline{F - D^n}$, and let $\rho : (G/PL)^F \longrightarrow (G/PL)^{F^*}$ be the restriction map. Then the composite $\rho\theta$ induces isomorphisms*

$$(\rho\theta)_* \; : \; \pi_r(G_F/\widetilde{PL}_F) \longrightarrow \pi_r((G/PL)^{F^*})$$

*for $r \geq 1$.*

**Remark.** For any based space $X$ let $X_0$ be the component of $X$ containing the base-point. Then $(G_F/\widetilde{PL}_F)_0$ is homotopy equivalent to $((G/PL)^{F^*})_0$, but $(G/PL)^{F^*}$ usually has more components than $G_F/\widetilde{PL}_F$.

**Proof.** First we prove that $(\rho\theta)_*$ is surjective; we defer the case $n = 4, r = 1$ until after Theorem 7. Let $B$ be a cell complex with $|B| = S^r$, and let $\beta$ be a principal cell of $B$. Let $f : S^r, \beta \longrightarrow (G/PL)^{F^*}, \text{bpt}$ represent an element of $x \in \pi_r((G/PL)^{F^*})$. Let

$$g \; : \; F^* \times S^r, F^* \times \beta \longrightarrow (G/PL), \text{bpt}$$

be the adjoint map. Extend $g$ over $(F^* \times S^r) \cup (F \times \beta)$ by defining $g(F \times \beta) = \text{bpt}$. Let $Q = F \times S^r$, $W = F \times \beta$ and let $Q^*$ be obtained from $Q$ by deleting the interior of an $(n+r)$-disc in $Q - W$. Then $Q^*$ deformation retracts onto $(F^* \times S^r) \cup (F \times \beta)$, so $g$ defines a homotopy class of maps $h : Q^*, W \longrightarrow G/PL, \text{bpt}$.

Let $k$ be large, identify $D^k$ with the northern hemisphere of $S^k$ and identify $2D^k$ with the closed region to the north of the Antarctic circle. Let $q : Q \longrightarrow S^k$ be an embedding such that $q^{-1}(D^k) = W$, $q^{-1}(2D^k) = Q^*$. Let $\nu_Q$ be a normal bundle of $Q$ in $S^k$ such that $\nu_Q|W$, $\nu_Q|Q^*$ are normal bundles of $W$, $Q^*$ in $D^k$, $2D^k$ respectively. Let $\phi^* : S^k \longrightarrow Q^{*\nu}/\partial Q^{*\nu}$ be the collapsing map.

Choose a piecewise linear bundle $\bar{\nu}_{Q^*}$ over $Q^*$ and a fibre homotopy equivalence $\bar{f} : \bar{\nu}_{Q^*} \longrightarrow \nu_{Q^*}$ such that $\bar{\nu}_{Q^*}|W = \nu_{Q^*}|W$, $\bar{f}$ is the identity on $\bar{\nu}_{Q^*}|W$ and $(\tau_{Q^*} \oplus \bar{\nu}_{Q^*}, 1 \oplus \bar{f})$ represents $h$. By the theorem of Wall quoted above, there is a map $\bar{\phi} : S^k \longrightarrow Q^{*\bar{\nu}}/\partial Q^{*\bar{\nu}}$ such that $\bar{f}\bar{\phi} \simeq \phi^*$. Let $\eta : Q^{*\bar{\nu}} \longrightarrow Q^{*\bar{\nu}}/\partial Q^{*\bar{\nu}}$ be the collapsing map; if $k$ is large enough then there is a map $\psi' : 2D^k, 2S^{k-1} \longrightarrow Q^{*\bar{\nu}}/\partial Q^{*\bar{\nu}}$ such that $\eta\psi'$ and $\bar{\phi}$ represent the same element of $\pi_k(Q^{*\bar{\nu}}/\partial Q^{*\bar{\nu}})$.

Adjust $\psi'$ by a homotopy until $\psi'|D^k = \phi|D^k$, and $\psi'$ is transverse regular on $Q^* \subset Q^{*\bar{\nu}}$; let $E' = \psi'^{-1}(Q^*)$, so $W \subset E'$. We shall modify $E', \partial E'$ by surgery (keeping $W$ fixed), attempting to make $\psi'| : E', \partial E' \longrightarrow Q^*, \partial Q^*$ a homotopy equivalence of pairs.

Since the inclusion induces an isomorphism $\pi_1(\partial Q^*) \longrightarrow \pi_1(\overline{Q^* - W})$ (in fact both groups are zero) and $n+r \geq 6$, we can use Theorem 3.3 of [17] to the manifold $\overline{E' - W}$. This has two boundary components, namely $\partial W$ and $\partial E'$; we wish to do surgery on $\mathrm{Int}(\overline{E' - W})$ and $\partial E'$, but **not** on $\partial W$.

We obtain a map $\psi^* : 2D^k, 2S^{k-1} \longrightarrow Q^{*\bar{\nu}}, \partial Q^{*\bar{\nu}}$, which is transverse regular on $Q^*$ and is homotopic to $\psi'$ (rel $D^k$), with the following property. Let

$$E^* = \psi^{*-1}(Q^*) ;$$

then

$$\psi^*| : \overline{E' - W}, \partial E^* \longrightarrow \overline{Q^* - W}, \partial Q^*$$

is a homotopy equivalence of pairs. It follows that $\psi^*| : E^*, \partial E^* \longrightarrow Q^*, \partial Q^*$ is a homotopy equivalence of pairs.

Since $\partial Q^* \cong S^{n+r-1}$ and $n + r - 1 \geq 5$, $\partial E^*$ is homeomorphic to $S^{n+r-1}$. Let $E = E^* \cup_{\partial E^*} D^{n+r}$, and extend the embedding $E^* \subset 2D^k$ to an embedding $E \subset S^k$. Let $\nu_E$ be a normal bundle of $E$ in $S^k$ such that $\nu_E|W$, $\nu_E|E^*$ are normal bundles of $W$, $E^*$ in $D^k$, $2D^k$ respectively. Extend $\psi^* : 2D^k \longrightarrow Q^{*\bar{\nu}}$ to a map $\psi : S^k \longrightarrow Q^{\bar{\nu}}$, transverse regular on $Q \subset Q^{\bar{\nu}}$ and with $E = \psi^{-1}(Q)$. Then $\psi|E \longrightarrow Q$ is a homotopy equivalence, and $\psi|W$ is the identity.

Recall that $B$ is a cell complex with $|B| = S^r$, and $\beta$ is a principal cell of $B$. Let $\gamma$ be an $(r-1)$-cell of $B$ contained in $\partial\beta$. Choose a $PL$ homeomorphism $k : |\partial\beta - \gamma| \times I \longrightarrow |B - \beta|$ such that $k_0$ is the inclusion. Recall that $Q = F \times |B|$. Since $n + r \geq 6$, we can use the relative $h$-cobordism theorem [12] to extend

$$\psi^{-1}(1 \times k)| : F \times |\partial\beta - \gamma| \times 0 \longrightarrow \partial\overline{E - W}$$

to a homeomorphism $H : F \times |\partial\beta - \gamma| \times I \longrightarrow \overline{E - W}$.

Define a block bundle $\xi$ over $B$ with $E(\xi) = E$ by $E_\beta(\xi) = W$ and, for each cell $\delta$ in $(B - \beta)$, $E_\delta(\xi) = H(1 \times k^{-1})(F \times \delta)$. Then $\xi$ satisfies the local triviality condition in the definition of a block bundle. Let

$$b(\xi) = 1 \times \mathrm{bpt} : F \longrightarrow F \times \mathrm{bpt} = E_{\mathrm{bpt}}(\xi) .$$

Let $t = p_1\psi : E \longrightarrow F$; then $(\xi, t)$ is a $G_F/\widetilde{PL}_F$-bundle over $S^r$, representing an element $y \in \pi_r(G_F/\widetilde{PL}_F)$. It is easily checked that $p_*(T(y)) \in \pi_r((G/PL)^{F^*})$ is represented by $(\tau_{Q^*} \oplus \bar{\nu}_{Q^*}, 1 \oplus \bar{f})$ so $p_*(T(y)) = x$. Therefore $(\rho\theta)_*(y) = x$, so $(\rho\theta)_*$ is surjective, as required (provided $n + r \geq 6$).

Similar arguments prove that $(\rho\theta)_*$ is injective; we have to consider $G_F/\widetilde{PL}_F$-bundles $(\xi_0, t_0)$, $(\xi_1, t_1)$ over $S^r \times 0$, $S^r \times 1$. We prove that they are isomorphic by extending them to a $G_F/\widetilde{PL}_F$-bundle $(\xi, t)$ over $S^r \times I$. Since $n + \dim(S^r \times I) \geq 6$, we can always carry out surgery and use the $h$-cobordism theorem. Thus the Theorem is established, except for surjectivity of $(\rho\theta)_*$ when $n = 4, r = 1$.

**Theorem 6**. *Let $F^n$ be a compact PL manifold with $\pi_1(\partial F)$ isomorphic to $\pi_1(F)$ by inclusion and $n \geq 6$. Then $\theta$ induces isomorphisms*

$$\theta_* \ : \ \pi_r(G_F/\widetilde{PL}_F) \longrightarrow \pi_r((G/PL)^F)$$

*for $r \geq 1$.*

**Proof.** Since the proof is essentially the same as the proof of Theorem 5, we shall not give the details. To prove that $\theta_*$ is surjective, let $B$, $\beta$, $\xi$, $Q$, $W$ be as above. Since $Q$ has a boundary $\partial Q$ such that $\pi_1(\partial Q) \longrightarrow \pi_1(\overline{Q - W})$ is an isomorphism, it is unnecessary to cut out a disc from $Q$. We can use Theorem 3.3 of [17] to construct a manifold $E \supset W$ with boundary $\partial E$ and a **simple** homotopy equivalence $\psi : E, \partial E \longrightarrow Q, \partial Q$ with $\psi | W$ equal to the identity.

In the construction of the block bundle $\xi$ above, we used the $h$-cobordism theorem to construct a homeomorphism $F \times |B - \beta| \longrightarrow \overline{E - W}$. Here we can use the $s$-cobordism theorem [7] twice (first for $\partial F \times |B - \beta|$, then for $F \times |B - \beta|$), since $\psi$ is a simple homotopy equivalence and $\dim(\partial F \times |B - \beta|) \geq 6$. The rest of the proof proceeds as above.

## IV. Periodicity of $G/PL$

In his thesis, Sullivan interpreted $[M, G/PL]$ in terms of $PL$ structures on manifolds homotopy equivalent to $M$. Thus it is useful to have information about $G/PL$ which facilitates computation of $[M, G/PL]$. It has been known for some time that $\pi_r(G/PL) \cong \mathbb{Z}, 0, \mathbb{Z}_2, 0$ according as $r \equiv 0, 1, 2, 3 \pmod 4$; in particular, $\pi_r(G/PL) \cong \pi_{r+4}(G/PL)$.

**Theorem 7**. *There is a map $\lambda : G/PL \longrightarrow \Omega^4(G/PL)$ such that $\lambda_* : \pi_r(G/PL) \longrightarrow \pi_{r+4}(G/PL)$ is an isomorphism if $r \neq 0, 4$ and a monomorphism onto a subgroup of index 2 if $r = 4$.*

**Proof.** Let $F^n$ be a closed 1-connected $PL$ manifold with $n \geq 4$. If $X$ is a based space we write $X_0$ for the component of $X$ containing the base-point. Consider the diagram

$$(1) \qquad \begin{array}{ccccc} & & (G_F/\widetilde{PL_F})_0 & & \\ & & \theta \swarrow \quad \searrow \rho\theta & & \\ \Omega^n(G/PL)_0 & \xrightarrow{\alpha} & (G/PL)_0^F & \xrightarrow{\rho} & (G/PL)_0^{F^*} \end{array}$$

where $\alpha$ is induced by a map $F \longrightarrow S^n$ of degree 1.

Suppose first that $n \geq 5$. Then $\rho\theta$ is a homotopy equivalence by Theorem 5 (we are not using the unproved case!). Let $\gamma'$ be a homotopy inverse to $\rho\theta$. Let $\gamma = \theta\gamma'$, so $\rho\gamma \simeq 1 : (G/PL)_0^{F^*} \longrightarrow (G/PL)_0^{F^*}$.

The Whitney sum construction gives a multiplication map $\mu : G/PL \times G/PL \longrightarrow G/PL$. If $K$ is a finite $CW$ complex, $\mu$ defines Abelian group structures on

$$[K, \Omega^n(G/PL)_0] \ , \ [K, (G/PL)_0^F] \ , \ [K, (G/PL)_0^{F^*}]$$

such that $\alpha_*$, $\rho_*$ are homeomorphisms. Let $x \in [K, (G/PL)_0^F]$ and let $y = (1 - \gamma_*\rho_*)(x)$, so $\rho_*(y) = 0$. Therefore $y = \alpha_*(z)$ for some $z \in [K, \Omega^n(G/PL)_0]$. Since $\rho$ has a right homotopy inverse, $\alpha_*$ is injective and $z$ is unique. Define a natural transformation

$$S : [\ , (G/PL)_0^F] \longrightarrow [\ , \Omega^n(G/PL)_0]$$

on finite $CW$ complexes by $S(x) = z$. By Lemma 1.7 of [3], there is a map

$$\sigma : (G/PL)_0^F \longrightarrow \Omega^n(G/PL)_0$$

with $S = \sigma_*$. Observe that

$$\alpha_*\sigma_*\alpha_* = \alpha_* - \gamma_*\rho_*\alpha_* = \alpha_* ,$$

since $\rho\alpha \simeq \text{bpt}$. Since $\alpha_*$ is injective, $\sigma_*\alpha_* = 1$.

Let $r \geq 1$ and consider the homomorphism

$$\sigma : \pi_r((G/PL)^F) \longrightarrow \pi_{n+r}(G/PL) .$$

This is an epimorphism (with right inverse $\alpha_*$). Let $x \in \pi_r((G/PL)^F)$ be represented by

$$g : F \times S^r, F \times \beta \longrightarrow G/PL, \text{bpt}$$

(where $\beta$ is a cell of $S^r$ containing the base-point). Let $Q = F \times S^r$, $W = F \times \beta$, as above.

Let $k$ be large and identify $D^k$ with the northern hemisphere of $S^k$. Let $q : Q \longrightarrow S^k$ be an embedding such that $q^{-1}(D^k) = W$. Let $\nu_Q$ be a normal bundle of $Q$ in $S^k$ such that $\nu_Q|W$ is a normal bundle of $W$ in $D^k$. Let $\phi : S^k \longrightarrow Q^\nu$ be the collapsing map.

Choose a piecewise linear bundle $\bar{\nu}_Q$ over $Q$ and a fibre homotopy equivalence $\bar{f} : \bar{\nu}_Q \longrightarrow \nu_Q$ such that $\bar{\nu}_Q|W = \nu_Q|W$, $\bar{f}$ is the identity on $\bar{\nu}_Q|W$ and $(\tau_Q \oplus \bar{\nu}_Q, 1 \oplus \bar{f})$ represents $g$. As in Chapter III, there is a map $\psi' : S^k \longrightarrow Q^{\bar{\nu}}$ such that $\bar{f}\psi' \simeq \phi$. Adjust $\psi'$ by a homotopy until $\psi'|D^k = \phi|D^k$ and $\psi'$ is transverse regular on $Q \subset Q^{\bar{\nu}}$; let $E' = \psi'^{-1}(Q)$, so $W \subset E'$. We attempt to modify $E'$ by surgery (keeping $W$ fixed), to make $\psi'|E' \longrightarrow Q$ a homotopy equivalence.

We seek a map $\psi : S^k \longrightarrow Q^{\bar{\nu}}$ which is transverse regular on $Q$ and is homotopic to $\psi'($ rel $D^k)$, and with the following property. Let $E = \psi^{-1}(Q)$; then $\psi|E \longrightarrow Q$ is a homotopy equivalence. Let $P_r = \mathbb{Z}, 0, \mathbb{Z}_2, 0$ according as $r \equiv 0, 1, 2, 3 \pmod 4$ (as in [8]). By [14,§4], since $Q$ is 1-connected and $\dim Q \geq 5$, there is an obstruction $\bar{\sigma}(x) \in P_{n+r}$ to performing the surgery. Note that $\bar{\sigma}(x)$ depends only on $x$.

Using the homotopy group addition in $\pi_r((G/PL)^F)$ (**not** the $H$-structure on $G/PL$) and the interpretation of $\bar{\sigma}(x)$ as a signature or Arf invariant, we see that $\bar{\sigma} : \pi_r((G/PL)^F) \longrightarrow P_{n+r}$ is a homomorphism. Consider the homomorphism $\bar{\sigma}\alpha_* : \pi_{n+r}(G/PL) \longrightarrow P_{n+r}$. This coincides with the canonical homomorphism obtained in [13], and is therefore an isomorphism. It follows that $\bar{\sigma}$ is an epimorphism.

If $x \in \pi_r((G/PL)^F)$, then $\gamma_*\rho_*(x)$ is represented by $\bar{g} : Q, W \longrightarrow G/PL$, bpt, where $\bar{g}$ agrees with $g$ on $Q^* = \overline{Q - D^{n+r}}$, and $\bar{g}|D^{n+r}$ is chosen so that $\bar{\sigma}(\bar{g}) = 0$ (because the surgery problem for $\gamma_*\rho_*(x) \in \mathrm{im}(\theta_*)$ is clearly soluble). Since $\bar{\sigma}\alpha_*$ is a monomorphism, these conditions characterise the homotopy class of $\bar{g}$. Therefore $x = \gamma_*\rho_*(x)$ if and only if $\bar{\sigma}(x) = 0$, so $\ker \sigma_* = \ker \bar{\sigma}$. If we identify $\pi_{n+r}(G/PL)$ with $P_{n+r}$ via the canonical isomorphism, we see that $\bar{\sigma}(x) = \sigma_*(x)$.

Let $\epsilon : G/PL \longrightarrow (G/PL)^F$ be induced by the map $F \longrightarrow$point, and let $\lambda^F$ denote the composite

$$G/PL \xrightarrow{\epsilon} (G/PL)^F \xrightarrow{\sigma} \Omega^n(G/PL) .$$

If $\dim F = n = 4$, this construction fails as

$$(\rho\theta)_* \; : \; \pi_1(G_F/\widetilde{PL_F}) \longrightarrow \pi_1((G/PL)^{F^*})$$

is not yet known to be surjective. However, we can construct a map $\bar{\lambda}^F : \Omega^4(G/PL)_0 \longrightarrow \Omega^{n+4}(G/PL)_0$; simply apply the functor $\Omega^4$ to diagram (1) and argue as above.

Let $x \in \pi_r(G/PL)$ be represented by $g : S^r \longrightarrow G/PL$. Then $\epsilon_*(x)$ is represented by $gp_2 : F \times S^r \longrightarrow G/PL$. Note that $x = \bar{\sigma}(x)$ and

$$\bar{\sigma}(\epsilon_*(x)) \;=\; \sigma_*\epsilon_*(x) \;=\; \lambda_*^F(x) .$$

Now $\bar{\sigma}(x)$ is the obstruction to making a certain map $\psi'| : V' \longrightarrow S^r$ a homotopy equivalence by surgery (where $V'$ is a certain framed $r$-manifold). Similarly $\bar{\sigma}(\epsilon_*(x))$ is the obstruction to making $1 \times \psi'| : F \times V' \longrightarrow F \times S^r$ a homotopy equivalence.

Take $F = \mathbb{CP}^2 \times \mathbb{CP}^2$. Suppose $r \equiv 0 \pmod 4$; then by [14], $\bar{\sigma}(x) = \frac{1}{8}(\text{signature of } V')$ if $r \geq 8$; but $\bar{\sigma}(x) = \frac{1}{16}(\text{signature of } V')$ if $r = 4$. Similarly,

$$\begin{aligned}\bar{\sigma}(\epsilon_*(x)) &= \tfrac{1}{8}(\text{signature of } F \times V' - \text{signature of } F \times S^r) \\ &= \tfrac{1}{8}(\text{signature of } V') \quad \text{for all } r.\end{aligned}$$

Thus $\bar{\sigma}(\epsilon_*(x)) = \bar{\sigma}(x)$ unless $r = 4$, when $\bar{\sigma}(\epsilon_*(x)) = 2\bar{\sigma}(x)$.

If $r \equiv 2 (\text{mod } 4)$, then it follows from Theorem 9.9 of [17] that $\bar{\sigma}(x) = \bar{\sigma}(\epsilon_*(x))$. (The theorem is stated for $r \geq 5$, but the argument seems to work when $r = 2$.) Since $\pi_r(G/PL) = \pi_{r+8}(G/PL) = 0$ if $r$ is odd, we have proved that $\lambda_*^F : \pi_r(G/PL) \longrightarrow \pi_{r+8}(G/PL)$ is an isomorphism if $r \neq 0, 4$, and a monomorphism onto a subgroup of index 2 if $r = 4$.

Similar arguments show that, if $F = \mathbb{CP}^2$ and $r \geq 1$, then $\bar{\lambda}_*^F : \pi_{r+4}(G/PL) \longrightarrow \pi_{r+8}(G/PL)$ is an isomorphism. Therefore $\bar{\lambda}^F : \Omega^4(G/PL)_0 \longrightarrow \Omega^8(G/PL)_0$ is a homotopy equivalence. Let $\lambda : G/PL \longrightarrow \Omega^4(G/PL)$ be the composite of $\lambda^{\mathbb{CP}^2 \times \mathbb{CP}^2}$ with a homotopy inverse to $\bar{\lambda}^{\mathbb{CP}^2}$; then $\lambda$ has the desired properties.

Now we can complete the proof of Theorem 5 by showing that, if $\dim F = 4$, then

$$(\rho\theta)_* \ : \ \pi_1(G_F/\widetilde{PL}_F) \longrightarrow \pi_1((G/PL)^{F^*})$$

is surjective. Consider the following diagram:

$$\begin{array}{ccccc}
\pi_1((G/PL)^F) & \xrightarrow{\rho_*} & \pi_1((G/PL)^{F^*}) & \xrightarrow{\partial} & \pi_0(\Omega^4(G/PL)) \\
\downarrow \lambda_* & & \downarrow \lambda_* & & \downarrow \lambda_* \\
\pi_1((\Omega^4(G/PL))^F) & \xrightarrow{\rho_*} & \pi_1((\Omega^4(G/PL))^{F^*}) & \xrightarrow{\partial} & \pi_0(\Omega^8(G/PL)) .
\end{array}$$

The rows are taken from the homotopy exact sequences of the Hurewicz fibrations

$$(G/PL)^F \longrightarrow (G/PL)^{F^*} \ , \ (\Omega^4(G/PL))^F \longrightarrow (\Omega^4(G/PL))^{F^*} \ .$$

The proof of Theorem 7 shows that, in the bottom row, $\rho_*$ is surjective so $\partial = 0$. But

$$\lambda_* \ : \ \pi_0(\Omega^4(G/PL)) \longrightarrow \pi_0(\Omega^8(G/PL))$$

is injective, so $\partial = 0$ in the top row. Therefore

$$\rho_* \ : \ \pi_1((G/PL)^F) \longrightarrow \pi_1((G/PL)^{F^*})$$

is surjective.

Let $x \in \pi_1((G/PL)^{F^*})$, and choose an element

$$\bar{x} \in \pi_1((G/PL)^F)$$

such that $\rho_*(\bar{x}) = x$. Let $\beta$ be an interval in $S^1$ containing the base-point, let $Q = F \times S^1$, $W = F \times \beta$. Let $\nu_Q$, $\psi'$ be as in the proof of Theorem 7. Since $\bar{\sigma}(x) \in P_5 = 0$, we can do surgery to find a map $\psi : S^k \longrightarrow Q^{\bar{\nu}}$ which is transverse regular to $\psi'$ (rel $D^k$), with the following property. Let $E = \psi^{-1}(Q)$; then $\psi| : E \longrightarrow Q$ is a homotopy equivalence.

Let $b_0, b_1$ be the end-points of $\beta$, and let $B$ be the cell complex $\{b_0, b_1, \beta, \overline{S^1 - \beta}\}$. Then $E - W$ is an $h$-cobordism between $F \times b_0$ and $F \times b_1$, and the $PL$ homeomorphism $1 \times b_1 : F \times b_0 \longrightarrow F \times b_1$ is in the preferred homotopy class. By Barden's $h$-cobordism theorem for 5-manifolds [2], there is a $PL$ homeomorphism $H : F \times |B - \beta| \longrightarrow \overline{E - W}$ with $H(F \times b_i) = F \times b_i$. Now we can define a block bundle $\xi$ over $B$ with $E(\xi) = E$, and a map $t : E \longrightarrow F$, as in the proof of Theorem 5. We obtain a $G_F/\widetilde{PL}_F$-bundle $(\xi, t)$ over $B$, representing an element $y \in \pi_1(G_F/\widetilde{PL}_F)$ such that $\theta_*(y) = \bar{x}$. Therefore $x = (\rho\theta)_*(y)$, so

$$(\rho\theta)_* \ : \ \pi_1(G_F/\widetilde{PL}_F) \longrightarrow \pi_1((G/PL)^{F^*})$$

is surjective. This completes the proof of Theorem 5.

## V. Topologically Trivial Block Bundles

Let $\xi$ be a block bundle over $B$ with fibre $F$. A **proper trivialisation** of $\xi$ is a proper map

$$h \ : \ E(\xi) \longrightarrow F \times |B|$$

such that

$$h(E_\beta(\xi)) \subset F \times \beta \quad \text{for each } \beta \in B$$

(base-points will be irrelevant in this chapter). Two proper trivialisations $h_0$, $h_1$ of $\xi$ are **properly homotopic** if there is a proper map

$$H \ : \ E(\xi) \times I \longrightarrow F \times |B|$$

such that

$$H(E_\beta(\xi) \times I) \subset F \times \beta$$

for each $\beta \in B$ and $H_t = h_t$ $(t = 0, 1)$. A **topological trivialisation** of $\xi$ is a proper trivialisation which is a topological homeomorphism; a $PL$ **trivialisation** is defined similarly.

**Theorem 8.** *Let $\xi$ be a block bundle over $B$ with fibre $\mathbb{R}^q$ ($q \geq 3$). Let $h : E(\xi) \longrightarrow \mathbb{R}^q \times |B|$ be a topological trivialisation of $\xi$. Then there is an obstruction $w \in H^3(B; \mathbb{Z}_2)$ which vanishes if and only if $h$ is properly homotopic to a PL trivialisation of $\xi$.*

**Proof.** Let $V$, $W$ be $PL$ manifolds and let $N$ be a compact submanifold of $W$ with $\partial N = N \cap \partial W$. A map $\phi : V \longrightarrow W$ is **$h$-regular** on $N$ if it is transverse regular on $N$ and $\phi| : \phi^{-1}(N) \longrightarrow N$ is a homotopy equivalence. Let $Q$ denote $\mathbb{C}\mathbb{P}^2 \times \mathbb{C}\mathbb{P}^2$. Our first objective is to construct the following:

(1) A proper map $f : E(\xi) \times Q \longrightarrow \mathbb{R}^q \times |B| \times Q$ such that, for each $\beta \in B$,

$$f| \ : \ E_\beta(\xi) \times Q \longrightarrow \mathbb{R}^q \times \beta \times Q$$

is $h$-regular on $0 \times \beta \times Q$.

(2) A proper homotopy $F$ from $h \times 1$ to $f$ such that, for each $\beta \in B$,

$$F(E_\beta(\xi) \times Q \times I) \subset \mathbb{R}^q \times \beta \times Q \ .$$

We shall eventually use $f$ and $F$ to construct a $PL$ trivialisation of $\xi$. The factor $Q$ is introduced to avoid difficulties with low-dimensional manifolds.

Let $T = \partial\Delta^2$ and write $T^r$ for the product of $r$ copies of $T$. Note that the universal covering space $\widetilde{T}^r$ of $T$ is $PL$ homeomorphic to $\mathbb{R}^r$. Choose a $PL$ embedding $\mathbb{R} \times T^{q-1} \subset \mathbb{R}^q$ and a $PL$ homeomorphism $\mathbb{R}^q \longrightarrow \mathbb{R} \times \widetilde{T}^{q-1}$ such that the composite

$$e \ : \ \mathbb{R}^q \longrightarrow \mathbb{R} \times \widetilde{T}^{q-1} \longrightarrow \mathbb{R} \times T^{q-1} \subset \mathbb{R}^q$$

is the identity on a neighbourhood of the origin.

Let $A$ be a subcomplex of $B$. Let $W_{A,r}$ denote $\mathbb{R}^r \times T^{q-r} \times |A| \times Q$ and let $N_{A,r} = 0 \times T^{q-r} \times A \times Q \subset W_{A,r}$. We have an embedding $W_{A,1} \subset \mathbb{R}^q \times A \times Q$ and there is a covering map $p : W_{A,r} \longrightarrow W_{A,r-1}$. Define $V_{A,1} = (h \times 1)^{-1}(W_{A,1})$ and let $g_{A,1} = h \times 1| : V_{A,1} \longrightarrow W_{A,1}$. Define $V_{B,r}$, $g_{B,r}$ ($r \geq 2$) inductively as follows. Let $p : V_{B,r} \longrightarrow V_{B,r-1}$ be the covering map induced from $p : W_{B,r} \longrightarrow W_{B,r-1}$ by the homeomorphism $g_{B,r-1} : V_{B,r-1} \longrightarrow W_{B,r-1}$. Let $g_{B,r} : V_{B,r} \longrightarrow W_{B,r}$ be a homeomorphism such that $pg_{B,r} = g_{B,r-1}p$. Finally let $V_{A,r} = p^{-1}(V_{A,r-1})$ and let $g_{A,r} = g_{B,r}|V_{A,r}$. We write $W_r^n$, $N_r^n$, $V_r^n$, $g_r^n$ for $W_{B^n,r}$, $N_{B^n,r}$, $V_{B^n,r}$, $g_{B^n,r}$ respectively, and abbreviate $W_{B,r}$, $N_{B,r}$, $V_{B,r}$, $g_{B,r}$ to $W_r$, $N_r$, $V_r$, $g_r$.

Suppose inductively that we have constructed the following, for some integer $n$:

(1) A proper map $f_1^{n-1} : V_1^{n-1} \longrightarrow W_1^{n-1}$ such that, for each $\beta \in B^{n-1}$, $f_1^{n-1}|V_{\beta,1} \longrightarrow W_{\beta,1}$ is $h$-regular on $N_{\beta,1}$.

(2) A proper homotopy $F_1^{n-1}$ from $g_1^{n-1}$ to $f_1^{n-1}$ such that, for each $\beta \in B^{n-1}$, $F_1^{n-1}(V_{\beta,1} \times I) \subset W_{\beta,1}$.

Suppose also that $f_1^{n-1}$, $F_1^{n-1}$ are extensions of $f_1^{n-2}$, $F_1^{n-2}$.

Now let $\beta \in B^n - B^{n-1}$. Let $f_{\partial\beta,1} = f_1^{n-1}|V_{\partial\beta,1}$ and let $F_{\partial\beta,1} = F_1^{n-1}|V_{\partial\beta,1} \times I$. The inductive hypothesis ensures that $f_{\partial\beta,1}$ is transverse regular on $N_{\partial\beta,1}$. Thus $M_{\partial\beta,1} = f_{\partial\beta,1}^{-1}(N_{\partial\beta,1})$ is a submanifold of $V_{\partial\beta,1}$ of codimension 1.

**Lemma 10.** *$f_{\partial\beta,1}$ is h-regular on $N_{\partial\beta,1}$.*

**Proof.** Let $B$ be a cell complex. A **blocked space** $E$ over $B$ consists of a topological space $E$ and, for each $\beta \in B$, a subspace $E_\beta$ of $E$ such that the following conditions are satisfied:

(1) $\{E_\beta : \beta \in B\}$ is a locally finite covering of $E$.
(2) If $\beta, \gamma \in B$, then $E_\beta \cap E_\gamma = \bigcup_{\delta \subset \beta\cap\gamma} E_\delta$.
(3) If $\beta$ is a face of $\gamma \in B$, then the inclusion $E_\beta \subset E_\gamma$ is a homotopy equivalence.
(4) If $\beta \in B$ and $E_{\partial\beta} = \bigcup_{\gamma\subset\partial\beta} E_\gamma$, then the pair $(E_\beta, E_{\partial\beta})$ has the absolute extension property.

If $E^{(1)}$, $E^{(2)}$ are blocked spaces over $B$, a **blocked equivalence** $\phi : E^{(1)} \longrightarrow E^{(2)}$ is a continuous map such that $\phi(E_\beta^{(1)}) \subset E_\beta^{(2)}$ and $\phi| : E_\beta^{(1)} \longrightarrow E_\beta^{(2)}$ is a homotopy equivalence for each $\beta \in B$. Observe that $M_{\partial\beta,1}$ and $N_{\partial\beta,1}$ are blocked spaces over $\partial\beta$, and $f_{\partial\beta,1}| : M_{\partial\beta,1} \longrightarrow N_{\partial\beta,1}$ is a blocked equivalence.

Suppose inductively that, if $E^{(1)}$, $E^{(2)}$ are blocked spaces over $B^{s-1}$, then any blocked equivalence $\phi : E^{(1)} \longrightarrow E^{(2)}$ is a homotopy equivalence. Now let $\phi : E^{(1)} \longrightarrow E^{(2)}$ be a blocked equivalence over $B^s$.

Let $C^{(i)} = \bigcup_{\beta\in B^{s-1}} E_\beta^{(i)}$ and let $D^{(i)}$, $\partial D^{(i)}$ be the disjoint unions of

$$\{E_\beta^{(i)} : \beta \in B^s - B^{s-1}\}, \quad \{E_{\partial\beta}^{(i)} : \beta \in B^s - B^{s-1}\}.$$

Then $\partial D^{(i)} \subset D^{(i)}$ and there are maps $\lambda^{(i)} : \partial D^{(i)} \longrightarrow C^{(i)}$ such that $E^{(i)} = C^{(i)} \cup_{\lambda^{(i)}} D^{(i)}$. By induction, $\phi : C^{(1)} \longrightarrow C^{(2)}$ is a homotopy equivalence.

Now $\phi$ defines a homotopy equivalence $\psi : D^{(1)} \longrightarrow D^{(2)}$ such that $\phi\lambda^{(1)} = \lambda^{(2)}\psi|\partial D^{(1)}$. By induction, $\psi|\partial D^{(1)} \longrightarrow \partial D^{(2)}$ is a homotopy equivalence. The pairs $(D^{(i)}, \partial D^{(i)})$ satisfy the absolute extension condition; using a result in homotopy theory we deduce that $\phi : E^{(1)} \longrightarrow E^{(2)}$ is a homotopy equivalence. By induction, any blocked equivalence over a finite-dimensional complex is a homotopy equivalence, and the Lemma follows.

Now the $PL$ manifold $V_{\beta,1}$ has two tame ends (for definition see [11]) with free Abelian fundamental groups. Since $M_{\partial\beta,1} \subset V_{\partial\beta,1}$ is a homotopy equivalence (by Lemma 10), $M_{\partial\beta,1}$ bounds collars of the ends of $V_{\partial\beta,1}$. Since $\dim V_{\beta,1} \geq 8$, we can apply Siebenmann's theorem [11,§5] to construct a compact submanifold $M_{\beta,1}$ of

$V_{\beta,1}$ with boundary $M_{\partial\beta,1}$ and such that $M_{\beta,1} \subset V_{\beta,1}$ is a homotopy equivalence. As in [16], we can extend $f_{\partial\beta,1}$ to $f_{\beta,1} : V_{\beta,1} \longrightarrow W_{\beta,1}$, transverse regular on $N_{\beta,1}$ and with $M_{\beta,1} = f_{\beta,1}^{-1}(N_{\beta,1})$. We can also extend $F_{\partial\beta,1}$ to a proper homotopy $F_{\beta,1}$ from $g_{\beta,1}$ to $f_{\beta,1}$.

Do this for all $n$-cells $\beta$ of $B$ to obtain extensions $f_1^n$, $F_1^n$ of $f_1^{n-1}$, $F_1^{n-1}$ satisfying the inductive hypotheses. This completes our induction on $n$; we have defined the following:

(1) A proper map $f_1 : V_1 \longrightarrow W_1$ such that for each $\beta \in B$, $f_1| : V_{\beta,1} \longrightarrow W_{\beta,1}$ is $h$-regular on $N_{\beta,1}$.

(2) A proper homotopy $F_1$ from $g_1$ to $f_1$ such that, for each $\beta \in B$, $F_1(V_{\beta,1} \times I) \subset W_{\beta,1}$.

Suppose inductively that we have defined the following, for some integer $r \geq 1$:

(1) A proper map $f_r : V_r \longrightarrow W_r$ such that for each $\beta \in B$, $f_r| : V_{\beta,r} \longrightarrow W_{\beta,r}$ is $h$-regular on $N_{\beta,r}$.

(2) A proper homotopy $F_r$ from $g_r$ to $f_r$ such that, for each $\beta \in B$, $F_r(V_{\beta,r} \times I) \subset W_{\beta,r}$.

Let

$$\widetilde{N}_r = 0 \times \mathbb{R} \times T^{q-r-1} \times |B| \times Q \subset W_{r+1} .$$

If $p : W_{r+1} \longrightarrow W_r$ is the covering map then $\widetilde{N}_r = p^{-1}(N_r)$. Lift $F^r$ to a proper homotopy $\widetilde{F}_r$ from $g_{r+1}$ to a map $\tilde{f}_r : V_{r+1} \longrightarrow W_{r+1}$. Let $\widetilde{M}_r = \tilde{f}_r^{-1}(\widetilde{N}_r)$ and let $M_r = f_r^{-1}(N_r)$. Since $p| : \widetilde{M}_r \longrightarrow M_r$ is a covering map and $f_r : M_r \longrightarrow N_r$ is a homotopy equivalence, $\tilde{f}_r| : \widetilde{M}_r \longrightarrow \widetilde{N}_r$ is a proper homotopy equivalence. Let $A$ be a subcomplex of $B$. Let $\widetilde{W}_{A,r} = p^{-1}(N_{A,r})$, $\tilde{f}_{A,r} = \tilde{f}_r|V_{A,r+1}$, $\widetilde{F}_{A,r} = \widetilde{F}_r|V_{A,r+1} \times I$, $\widetilde{M}_{A,r} = \widetilde{M}_r \cap V_{A,r+1}$ and $M_{A,r} = M_r \cap V_{A,r}$.

We construct the following:

(1) A proper map $\phi_r : \widetilde{M}_r \longrightarrow \widetilde{N}_r$ such that for each $\beta \in B$, $\phi_r| : \widetilde{M}_{\beta,r} \longrightarrow \widetilde{N}_{\beta,r}$ is $h$-regular on $N_{\beta,r+1}$.

(2) A proper homotopy $\Phi_r$ from $\tilde{f}_r|\widetilde{M}_r$ to $\phi_r$ such that, for each $\beta \in B$, $\Phi_r(\widetilde{M}_{\beta,r} \times I) \subset \widetilde{N}_{\beta,r}$.

The construction is exactly the same as the one given above for $f_1$ and $F_1$. We apply Siebenmann's theorem to $\widetilde{M}_{\beta,r}$ instead of $V_{\beta,1}$; the details will be omitted.

Using the product structure on a neighbourhood of $\widetilde{M}_r$ in $V_{r+1}$, we can construct the following:

(1) A proper map $f_{r+1} : V_{r+1} \longrightarrow W_{r+1}$ such that for each $\beta \in B$, $f_{r+1}| : V_{\beta,r+1} \longrightarrow W_{\beta,r+1}$ is $h$-regular on $N_{\beta,r+1}$.

(2) A proper homotopy $F_{r+1}$ from $g_{r+1}$ to $f_{r+1}$ such that, for each $\beta \in B$, $F_{r+1}(V_{\beta,r+1} \times I) \subset W_{\beta,r+1}$.

This completes the induction on $r$. When $r = q$ we obtain a proper map $f_q : V_q \longrightarrow W_q = \mathbb{R}^q \times |B| \times Q$ and a proper homotopy $F_q$ from $g_q$ to $f_q$, satisfying the inductive hypotheses.

Consider the commutative diagram:

$$\begin{array}{ccccc} V_q & \xrightarrow{g_q} & W_q & = & \mathbb{R}^q \times |B| \times Q \\ \downarrow \epsilon & & \downarrow \epsilon & & \downarrow e \times 1 \\ E(\xi) \times Q & \xrightarrow{h \times 1} & \mathbb{R}^q \times |B| \times Q & = & \mathbb{R}^q \times |B| \times Q \end{array}$$

where $\epsilon$ denotes a covering map followed by an inclusion. Recall that $e : \mathbb{R}^q \longrightarrow \mathbb{R}^q$ is the identity on an open disc neighbourhood $U$ of the origin.

Let $A$ be a subcomplex of $B$, let $X_A$ denote

$$h^{-1}(U \times |A|) \times Q - h^{-1}(0 \times |A|) \times Q \subset E(\xi) \times Q ,$$

and let $X = X_B$, $X^n = X_{B^n}$. Suppose inductively that we have constructed the following, for some integer $n$.

(1) A subset $Y^{n-1}$ of $X^{n-1}$ such that, for each $\beta \in B^{n-1}$, $Y_\beta = Y^{n-1} \cap X_\beta$ is a compact submanifold of $X_\beta$ of codimension one and $Y_\beta \subset X_\beta$ is a homotopy equivalence. Then $E(\xi|B^{n-1}) \times Q - Y^{n-1}$ has two components; let $Z^{n-1}$ be the closure of the bounded component. Let $(Z')^{n-1}$ be the component of $\epsilon^{-1}(Z^{n-1})$ which lies in $g_q^{-1}(U \times |B^{n-1}| \times Q)$, and let $(Y')^{n-1} = (Z')^{n-1} \cap \epsilon^{n-1}(Y^{n-1})$.

(2) $PL$ homeomorphisms

$$\gamma^{n-1} : Y^{n-1} \times [0,\infty) \longrightarrow \overline{E(\xi|B^{n-1}) \times Q - Z^{n-1}} ,$$

$$(\gamma')^{n-1} : (Y')^{n-1} \times [0,\infty) \longrightarrow \overline{V_q^{n-1} - (Z')^{n-1}}$$

such that $\gamma_0^{n-1}$, $(\gamma')_0^{n-1}$ are the inclusions.

Suppose further that $\gamma^{n-1}$, $(\gamma')^{n-1}$ are extensions of $\gamma^{n-2}$, $(\gamma')^{n-2}$.

Now let $\beta \in B^n - B^{n-1}$. Let

$$\begin{aligned} Y_{\partial\beta} &= Y^{n-1} \cap X_{\partial\beta} , \\ \gamma_{\partial\beta} &= \gamma^{n-1}|Y_{\partial\beta} \times [0,\infty) , \\ \gamma'_{\partial\beta} &= (\gamma')^{n-1}|Y'_{\partial\beta} \times [0,\infty) . \end{aligned}$$

Then $Y_{\partial\beta}$ bounds a collar of the end of $E(\xi|\partial\beta) \times Q$. It follows that $Y_{\partial\beta} \subset X_{\partial\beta}$ is a homotopy equivalence; since $\dim X_{\partial\beta} \geq 8$, $Y_{\partial\beta}$ bounds a collar of the ends of

$X_{\partial\beta}$.

Since the ends of $X_\beta$ are tame and have trivial fundamental groups, Siebenmann's theorem shows that there is a compact submanifold $Y_\beta$ of $X_\beta$ with boundary $Y_{\partial\beta}$ and such that $Y_\beta \subset X_\beta$ is a homotopy equivalence. It follows that $Y_\beta$ bounds a collar of the end of $E(\xi|\beta) \times Q$. Let

$$\gamma_\beta \; : \; Y_\beta \times [0,\infty) \longrightarrow \overline{E(\xi|\beta) \times Q - Z_\beta}$$

be a $PL$ homeomorphism such that $(\gamma_\beta)_0$ is the inclusion and $\gamma_\beta|Y_{\partial\beta} \times [0,\infty) = \gamma_{\partial\beta}$. Do this for all $n$-cells $\beta$ of $B$ to obtain $Y^n$, $\gamma^n$ satisfying the inductive hypotheses.

Define $(Z')^n$, $(Y')^n$ as in (1) above, and note that $\epsilon : (Z')^n \longrightarrow Z^n$ is a $PL$ homeomorphism. Then, for each $\beta \in B^n - B^{n-1}$, $Y'_\beta \subset \overline{V_{\beta,q} - Z'_\beta}$ is a homotopy equivalence, so $Y'_\beta$ bounds a collar of the end of $V_{\beta,q}$. Let

$$\gamma'_\beta \; : \; Y'_\beta \times [0,\infty) \longrightarrow \overline{V_{\beta,q} - Z'_\beta}$$

be a $PL$ homeomorphism such that $(\gamma'_\beta)_0$ is the inclusion and $\gamma'_\beta|Y'_{\partial\beta} \times [0,\infty) = \gamma'_{\partial\beta}$. Then the $\gamma'_\beta$ fit together to define an extension $(\gamma')^n$ of $(\gamma')^{n-1}$ satisfying the inductive hypotheses. This completes the induction on $n$.

Let

$$Y = \bigcup_{n=1}^{\infty} Y^n \; , \quad Z = \bigcup_{n=1}^{\infty} Z^n \; , \quad \gamma = \bigcup_{n=1}^{\infty} \gamma^n \; , \quad \gamma' = \bigcup_{n=1}^{\infty} (\gamma')^n \; .$$

Define a $PL$ homeomorphism $\psi : E(\xi) \times Q \longrightarrow V_q$ by $\psi = \epsilon^{-1}$ on $Z$ and $\psi = \gamma^{-1}(\epsilon^{-1} \times 1)\gamma$ elsewhere. Define a proper homotopy $\Psi$ from $g_q\psi$ to $h \times 1$ as follows. If $x \in \mathbb{R}^q$, $y \in |B|$, $z \in Q$ and $t \in [0,1)$, let $g_q\psi(h^{-1}(tx,y),z) = (x',y',z')$, and define

$$\Psi(h^{-1}(x,y),z,t) = (t^{-1}x',y',z') \; .$$

Define

$$\Psi(h^{-1}(x,y),z,0) = (x,y,z) \; ;$$

this makes $\Psi$ continuous since $(x',y',z') = (tx,y,z)$ provided $t$ is sufficiently small.

Now we can define the proper map $f : E(\xi) \times Q \longrightarrow \mathbb{R} \times |B| \times Q$ and proper homotopy $F$ from $h \times 1$ to $f$, as promised at the beginning of the proof. Let $f = f_q\psi$ and let $F = \Psi * (F_q\psi)$ be defined by

$$F(x,t) = \begin{cases} \Psi(x,2t) & 0 \le t \le \frac{1}{2} \\ F_q\psi(x,2t-1) & \frac{1}{2} \le t \le 1 \; . \end{cases}$$

Then $f$ and $F$ have the required properties (1) and (2).

Suppose inductively that we have constructed the following, for some integer

$n$.

(1) A proper trivialisation $j^{n-1} : E(\xi|B^{n-1}) \longrightarrow \mathbb{R}^q \times |B^{n-1}|$.
(2) A proper homotopy $J^{n-1}$ from $h^{n-1}$ to $j^{n-1}$.
(3) A proper homotopy $L^{n-1}$ from $f^{n-1}$ to $j^{n-1} \times 1$ such that, for each $\beta \in B^{n-1}$, $L^{n-1}|E(\xi|\beta) \times Q$ is $h$-regular on $0 \times \beta \times Q$.
(4) A proper homotopy $\mathcal{L}^{n-1}$ from $\bar{F}^{n-1} * (J^{n-1} \times 1)$ to $L^{n-1}$ (rel $\mathbb{R}^q \times |B^{n-1}| \times Q \times \partial I$).

Suppose further that $j^{n-1}$, $J^{n-1}$, $L^{n-1}$, $\mathcal{L}^{n-1}$ are extensions of $j^{n-2}$, $J^{n-2}$, $L^{n-2}$, $\mathcal{L}^{n-2}$ respectively.

Let $\beta \in B^n - B^{n-1}$. If $A$ is a subcomplex of $B^{n-1}$, then $j_A$, $J_A$, $L_A$, $\mathcal{L}_A$ will have the usual meanings. As in Lemma 10 we see that

$$L_{\partial\beta} \; : \; E(\xi|\partial\beta) \times Q \times I \longrightarrow \mathbb{R}^q \times \partial\beta \times Q$$

is $h$-regular on $0 \times \partial\beta \times Q$. Note that $L_{\partial\beta}$ is a proper homotopy from $f_{\partial\beta}$ to $j_{\partial\beta} \times 1$. Extend $L_{\partial\beta}$ to a proper homotopy $K_\beta$ from $f_\beta$ to a proper map $k_\beta : E_\beta(\xi) \times Q \longrightarrow \mathbb{R}^q \times \beta \times Q$. We can arrange for $K_\beta$ to be $h$-regular on $0 \times \beta \times Q$.

Now $J_{\partial\beta}$ is a proper homotopy from $h_{\partial\beta}$ to $j_{\partial\beta}$. Extend $J_{\partial\beta}$ to a proper map $I_\beta$ from $h_\beta$ to a proper map $i_\beta : E_\beta(\xi) \longrightarrow \mathbb{R} \times \beta$. Using the homotopies $(I_\beta \times 1) * F_\beta * K_\beta$ and $\mathcal{L}_{\partial\beta}$, we see that $i_\beta \times 1$ is properly homotopic (rel $\mathbb{R}^q \times \partial\beta \times Q$) to $k_\beta$.

The obstruction to deforming $i_\beta$ properly (rel $E(\xi|\partial\beta)$) to a $PL$ homeomorphism $j'_\beta : E_\beta(\xi) \longrightarrow \mathbb{R}^q \times \beta$ is an element $x \in \pi_n(G/PL)$. Let $\lambda_* : \pi_n(G/PL) \longrightarrow \pi_{n+8}(G/PL)$ be the periodicity homomorphism discussed in Chapter IV. Then $\lambda_*(x)$ is the obstruction to deforming $i_\beta \times 1$ properly (rel $E(\xi|\partial\beta) \times Q$) to a map $k'_\beta$ which is $h$-regular on $0 \times \beta \times Q$. The previous paragraph shows that $\lambda_*(x) = 0$; since $\lambda_*$ is a monomorphism, $x = 0$. Choose a $PL$ homeomorphism $j'_\beta : E_\beta(\xi) \longrightarrow \mathbb{R} \times \beta$ and a proper homotopy $J'_\beta$ from $h_\beta$ to $j'_\beta$ extending $J_{\partial\beta}$.

Now $\mathcal{L}_{\partial\beta}$ is a proper homotopy from $\bar{F}_{\partial\beta} * (J_{\partial\beta} \times 1)$ to $L_{\partial\beta}$. Extend $\mathcal{L}_{\partial\beta}$ to a proper homotopy $\mathcal{G}_\beta$ (rel $E_\beta(\xi) \times Q \times \partial I$) from $\bar{F}_\beta * (J'_\beta \times 1)$ to a proper homotopy $G_\beta$ between $f_\beta$ and $j'_\beta \times 1$. Let $y \in \pi_{n+9}(G/PL)$ be the obstruction to deforming $G_\beta$ properly (rel $\partial(\mathbb{R}^q \times \beta \times Q \times I)$) to a homotopy $G'_\beta$ which is $h$-regular on $0 \times \beta \times Q$.

If we vary $(j'_\beta, J'_\beta)$ by an element $z \in \pi_{n+1}(G/PL)$, we replace $y$ by $y + \lambda_*(z)$. If $n \neq 3$ then $\lambda_*$ is surjective, so we can choose $z$ so that $y + \lambda_*(z) = 0$. In other words, we can replace $(j'_\beta, J'_\beta)$ by a pair $(j_\beta, J_\beta)$ for which $y$ vanishes. Then there is a proper homotopy $L_\beta$ from $f_\beta$ to $j_\beta \times 1$ which is $h$-regular on $0 \times \beta \times Q$, and a proper homotopy $\mathcal{L}_\beta$ (rel $\mathbb{R}^q \times \beta \times Q \times \partial I$) from $\bar{F}_\beta * (J_\beta \times 1)$ to $L_\beta$; $L_\beta$ and $\mathcal{L}_\beta$ are extensions of $L_{\partial\beta}$ and $\mathcal{L}_{\partial\beta}$ respectively.

Do this for all $n$-cells $\beta$ of $B$ to obtain $j^n$, $J^n$, $L^n$, $\mathcal{L}^n$ satisfying conditions (1)–(4). This completes the induction provided $n \neq 3$. In case $\beta$ is a 3-cell of $B$, let $c(\beta) \in \mathbb{Z}_2$ be the mod 2 reduction of $y \in \pi_{12}(G/PL) = \mathbb{Z}$. This defines a cochain $c \in C^3(B;\mathbb{Z}_2)$. The above argument enables us to construct $j^3$, $J^3$, $L^3$, $\mathcal{L}^3$ provided $c = 0$.

We consider the effect of varying $L^2$. Suppose $j^1$, $J^1$, $L^1$, $\mathcal{L}^1$, $j^2$, $J^2$, $L^2$, $\mathcal{L}^2$ are constructed, and let $\beta$ be a 3-cell in $B$. Observe that, if the cells $\alpha \subset \partial\beta$ are oriented suitably, then $\partial\beta = \sum_{\alpha\subset\partial\beta} \alpha \in C_2(B;\mathbb{Z})$. If we vary $L_\alpha$ by an element $u_\alpha \in \pi_{12}(G/PL) = \mathbb{Z}$, it can be seen that $c(\beta)$ is replaced by $c(\beta) + (\sum_{\alpha\subset\partial\beta} u_\alpha)_2$. Let $u \in C^2(B;\mathbb{Z}_2)$ be the cochain defined by $u(\alpha) = u_\alpha$; then we have replaced $c$ by $c + \delta u$.

Now let $\gamma$ be a 4-cell of $B$, so $\partial\gamma = \sum_{\beta\subset\partial\gamma} \beta \in C_3(B;\mathbb{Z})$. For each 3-cell $\beta \subset \partial\gamma$, define $j'_\beta$, $J'_\beta$ as above, and define

$$J'_{\partial\gamma} : E(\xi|\partial\gamma) \times I \longrightarrow \mathbb{R}^q \times \partial\gamma$$

by $J'_{\partial\gamma}|E(\xi|\beta) \times I = J'_\beta$. It is easy to adjust $(j'_\beta, J'_\beta)$ on one cell $\beta \subset \partial\gamma$ until $J'_{\partial\gamma}$ extends to a proper homotopy $J'_\gamma$ from $h_\gamma$ to a $PL$ homeomorphism $j'_\gamma$ : $E(\xi|\partial\gamma) \longrightarrow \mathbb{R}^q \times \partial\gamma$.

Define

$$G_{\partial\gamma} : E(\xi|\partial\gamma) \times Q \times I \longrightarrow \mathbb{R}^q \times \partial\gamma \times Q$$

by $G_{\partial\gamma}|E(\xi|\beta) \times Q \times I = G_\beta$. Let $v_\gamma \in \pi_{12}(G/PL) = \mathbb{Z}$ be the obstruction to deforming $G_{\partial\gamma}$ properly (rel $E(\xi|\partial\gamma) \times Q \times \partial I$) to a proper homotopy $G'$ which is $h$-regular on $0 \times \partial\gamma \times Q$. Then it can be seen that $v_\gamma = \sum_{\beta\subset\partial\gamma} y_\beta$, so $(\delta c)(\gamma) = c(\partial\gamma)$ is equal to the mod 2 reduction of $v_\gamma$.

On the other hand, $v_\gamma$ is the obstruction to deforming $\bar{F}_{\partial\gamma} * (J'_{\partial\gamma} \times 1)$ properly (rel $E(\xi|\partial\gamma) \times Q \times \partial I$) to a proper homotopy $G'_{\partial\gamma}$ which is $h$-regular on $0 \times \partial\gamma \times Q$. But $\bar{F}_{\partial\gamma} * (J'_{\partial\gamma} \times 1)$ extends to a proper homotopy $\bar{F}_\gamma * (J'_\gamma \times 1)$ from $f_\gamma$ to $j'_\gamma \times 1$, both of which are $h$-regular on $0 \times \gamma \times Q$. Now it follows from Wall's surgery theorem [14] that $v_\gamma = 0$. Therefore $(\delta c)(\gamma) = 0$, so $c$ is a cocycle.

Let $w \in H^3(B;\mathbb{Z}_2)$ be the cohomology class of $c$; we have shown that our construction can be carried out if $w = 0$. Assume now that $w = 0$, and let $j = \bigcup_{n=1}^{\infty} j^n$, $J = \bigcup_{n=1}^{\infty} J^n$. Then $j$ is a $PL$ trivialisation of $\xi$ and $J$ is a proper homotopy from $h$ to $j$, as required. It is not hard to see that $w = 0$ whenever $h$ is properly homotopic to a $PL$ trivialisation, so Theorem 8 is proved.

Theorem 8 implies a result on the Hauptvermutung by fairly well-known arguments, given in Sullivan's thesis.

**Corollary.** *Let $M^n$, $N^n$ be closed,* 1*-connected PL manifolds with* $n \geq 5$ *and let* $h : M \longrightarrow N$ *be a topological homeomorphism. If* $H^3(M; \mathbb{Z}_2) = 0$*, then* $h$ *is homotopic to a PL homeomorphism.*

## References


[1] M. F. Atiyah, **Thom complexes**, Proc. London Math. Soc. 11, 291–310 (1961)

[2] D. Barden, to appear.

[3] E. H. Brown, Jr. **Cohomology theories**, Ann. of Math. 75, 467–484 (1962)

[4] E. H. Brown, Jr. **Abstract homotopy theory**, Trans. Amer. Math. Soc. 119, 79–85 (1965)

[5] A. Dold, **Partitions of unity in the theory of fibrations**, Ann. of Math. 78, 223–255 (1963)

[6] A. Haefliger and C. T. C. Wall, **Piecewise linear bundles in the stable range**, Topology 4, 209–214 (1965)

[7] M. A. Kervaire, **Le théorème de Barden-Mazur-Stallings**, Comm. Math. Helv. 40, 31–42 (1965)

[8] J. Levine, **A classification of differentiable knots**, Ann. of Math. 82, 15–50 (1965)

[9] J. W. Milnor, **On spaces having the homotopy type of a CW-complex**, Trans. Amer. Math. Soc. 90, 272–280 (1959)

[10] C. P. Rourke and B. J. Sanderson, **Block Bundles I**, Ann. of Math. 87, 1–28 (1968)

[11] L. C. Siebenmann, **The obstruction to finding a boundary for an open manifold of dimension greater than five**, Thesis, Princeton University, 1965.

[12] S. Smale, **On the structure of manifolds**, Amer. J. Math. 84, 387–399 (1962)

[13] D. P. Sullivan, **Triangulating homotopy equivalences**, notes, Warwick University, 1966.

[14] C. T. C. Wall, **An extension of results of Novikov and Browder**, Amer. J. Math. 88, 20–32 (1966)

[15] C. T. C. Wall, **Poincaré complexes**, Ann. of Math. 86, 213–245 (1970)

[16] C. T. C. Wall, **On bundles over a sphere with fibre Euclidean space**, Fund. Math. 61, 57–72 (1967)

[17] C. T. C. Wall, **Surgery on compact manifolds** Academic Press (1970)

[18] J. H. C. Whitehead, **Combinatorial homotopy I**, Bull. Amer. Math. Soc. 55, 213–245 (1949)

[19] E. C. Zeeman, **Seminar on combinatorial topology III**, Inst. Hautes Etudes Sci. Paris, 1963.


University of Liverpool
Summer 1967

# Triangulating and Smoothing Homotopy Equivalences and Homeomorphisms. Geometric Topology Seminar Notes

By D. P. Sullivan

## Introduction

We will study the smooth and piecewise linear manifolds within a given homotopy equivalence class. In the first part we find an obstruction theory for deforming a homotopy equivalence between manifolds to a diffeomorphism or a piecewise linear homeomorphism. In the second part we analyze the piecewise linear case and characterize the obstructions in terms of a geometric property of the homotopy equivalence. In the third part we apply this analysis to the Hauptvermutung and complex projective space.

## I. Triangulating and Smoothing Homotopy Equivalences

**Definition 1.** Let $A_i$ denote the Abelian group of almost framed[1] cobordism classes of almost framed smooth $i$-manifolds.
Let $P_i$ denote the Abelian group of almost framed cobordism classes of almost framed piecewise linear $i$-manifolds.

**Theorem 1.** (**The obstruction theories**) *Let $f : (L, \partial L) \longrightarrow (M, \partial M)$ be a homotopy equivalence between connected piecewise linear $n$-manifolds. Let $Q$ be an $(n-1)$-dimensional submanifold of $\partial L$ such that $f(\partial L - Q) \subseteq \partial M - f(Q)$. Suppose that $n \geq 6$ and that $\pi_1(L) = \pi_1(\text{each component of } \partial L - Q) = 0$.*

(a) *If $f|Q$ is a PL-homeomorphism, then $f$ may be deformed (mod $Q$) to a PL-homeomorphism on all of $L$ iff a sequence of obstructions in $H^i(L, Q; P_i)$ $0 < i < n$ vanish.*

(b) *If $L$ and $M$ are smooth, $f|Q$ is a diffeomorphism, and $\partial L \neq Q$ then $f$ may be deformed (mod $Q$) to a diffeomorphism on all of $L$ iff a sequence of obstructions in $H^i(L, Q; A_i)$ $0 < i < n$ vanish.*

**Remark.** From the work of Kervaire and Milnor [KM] we can say the following about the above coefficient groups:

[1] almost framed means framed over some $(i-1)$-skeleton

(i) If $\theta_i$ denotes the finite Abelian group of oriented equivalence classes of differentiable structures on $S^i$, then there is a natural exact sequence

$$\ldots \longrightarrow P_{i+1} \xrightarrow{\partial} \theta_i \xrightarrow{i} A_i \xrightarrow{j} P_i \longrightarrow \theta_{i-1} \longrightarrow \ldots$$

with image$(\partial) = \theta_i\partial\pi = \{\pi\text{-boundaries}\} \subseteq \theta_i$.

(ii) $P_* = P_1, P_2, P_3, \ldots, P_i, \ldots$ is just the (period four) sequence

| $i$ | 1 | 2 | 3 | 4 | 5 | 6 | 7 | 8 |
|---|---|---|---|---|---|---|---|---|
| $P_i$ | 0 | $\mathbb{Z}_2$ | 0 | $\mathbb{Z}$ | 0 | $\mathbb{Z}_2$ | 0 | $\mathbb{Z}$ |

| 9 | 10 | 11 | 12 | 13 | 14 | 15 | 16 |
|---|---|---|---|---|---|---|---|
| 0 | $\mathbb{Z}_2$ | 0 | $\mathbb{Z}$ | 0 | $\mathbb{Z}_2$ | 0 | $\mathbb{Z}$ |

(iii) For $i \leq 19$, $A_i$ may be calculated

| $i$ | 1 | 2 | 3 | 4 | 5 | 6 | 7 |
|---|---|---|---|---|---|---|---|
| $A_i$ | 0 | $\mathbb{Z}_2$ | 0 | $\mathbb{Z}$ | 0 | $\mathbb{Z}_2$ | 0 |

| 8 | 9 | 10 | 11 | 12 | 13 |
|---|---|---|---|---|---|
| $\mathbb{Z} \oplus \mathbb{Z}_2$ | $2\mathbb{Z}_2$ | $\mathbb{Z}_6$ | 0 | $\mathbb{Z}$ | $\mathbb{Z}_3$ |

| 14 | 15 | 16 | 17 | 18 | 19 |
|---|---|---|---|---|---|
| $2\mathbb{Z}_2$ | $\mathbb{Z}_2$ | $\mathbb{Z} \oplus \mathbb{Z}_2$ | $3\mathbb{Z}_2$ | $\mathbb{Z}_2 \oplus \mathbb{Z}_8$ | $\mathbb{Z}_2$ |

Note that Theorem 1 is analogous to a fundamental theorem in smoothing theory. In that case $f$ is a $PL$-homeomorphism, $f|Q$ is a diffeomorphism, and $f$ may be deformed by a weak-isotopy (mod $Q$) to a diffeomorphism iff a sequence of obstructions in $H^i(L, Q; \theta_i)$ vanish.

These three obstruction theories are related by the exact sequence of coefficients above.

**Proof of Theorem 1**: There are several approaches to Theorem 1. The most direct method seeks to alter $f$ by a homotopy so that it becomes a diffeomorphism or a $PL$-homeomorphism on a larger and larger region containing $Q$. Suppose for example that $M$ is obtained from $f(Q)$ by attaching one $i$-handle with core disk $D_i$.

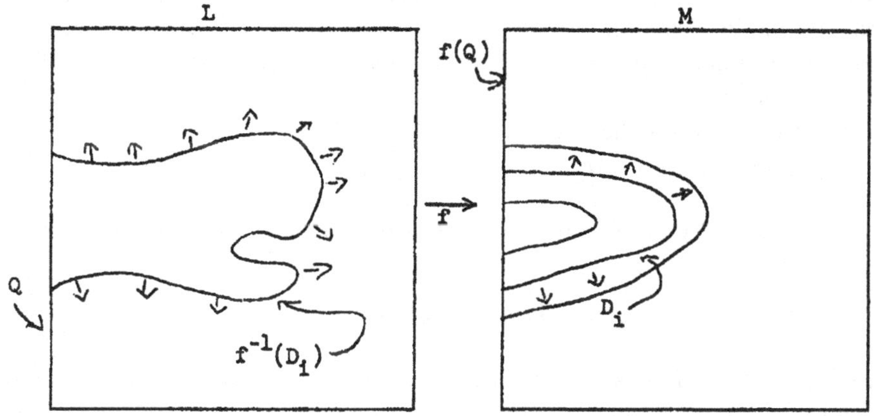


Figure 1

Then $f$ is deformed (mod $Q$) so that it is transverse regular to the framed manifold $D_i$. The framed manifold $f^{-1}(D_i)$ has a (smooth or $PL$) sphere boundary and determines an element in $A_i$ or $P_i$. If this element is zero, then surgery techniques may be employed to deform $f$ so that it is a diffeomorphism or $PL$-homeomorphism on a neighborhood of $f^{-1}D_i$.

Theorem 1 asserts that the cochain with values in $A_i$ or $P_i$ determined by the $f^{-1}D_i$'s has the properties of an obstruction cochain.

A complete description is given in [S1]. See also [W1].

The obstruction theories of Theorem 1 have the usual complications of an Eilenberg-Whitney obstruction theory. The $k^{\text{th}}$ obstruction in $H^k(L,Q;P_k)$ or $H^k(L,Q;A_k)$ is defined only when the lower obstructions are zero; and its value depends on the nature of the deformation of $f$ to a $PL$-homeomorphism or diffeomorphism on a thickened region of $L$ containing the $(k-2)$-skeleton of $L-Q$.

Thus applications of a theory in this form usually treat only the first obstruction or the case when the appropriate cohomology groups are zero.

For more vigorous applications of the theory one needs to know more precisely how the obstructions depend on the homotopy equivalence $f$ — for example, is it possible to describe the higher obstructions and their indeterminacies in terms of a priori information about $f$?[1]

[1] I am indebted to Professor Steenrod for suggesting this problem at my Thesis Defense, January 1966.

We will concentrate on the $PL$ obstruction theory – where a complete analysis can be made. We will replace the sequence of conditions in Theorem 1 by one condition which depends only on the geometrical invariants of $f$ (Theorem 2). These geometrical invariants are the classical surgery obstructions in $P_*$ obtained by studying the behavior of $f$ on the inverse image of certain characteristic (singular) submanifolds of $M$.

It would be very interesting if a similar analysis can be made of the smooth theories.

## The $PL$ Theory

**Definition 2.** Let $M$ be an oriented $PL$ $m$-manifold whose oriented boundary is the disjoint union of $n$ copies of the closed oriented $(m-1)$-manifold $L$. We call the polyhedron $V$ obtained from $M$ by identifying the copies of $L$ to one another a $\mathbb{Z}_n$**-manifold**. We denote the subcomplex $L \subseteq V$ by $\delta V$, the **Bockstein** of $V$.

A finite disjoint union of $\mathbb{Z}_n$-manifolds for various $n$'s and of various dimensions is called a **variety**.

If $X$ is a polyhedron, a **singular variety in** $X$ is a piecewise linear map $f : V \longrightarrow X$ of a variety $V$ into $X$.

**Remark.** (i) Note that if $V$ is a $\mathbb{Z}_n$-manifold of dimension $m$ then $V$ is locally Euclidean except along points of $\delta V = L$. A neighborhood of $L$ in $V$ is $PL$-homeomorphic to $L \times \text{cone}\,(n \text{ points})$.
(ii) A $\mathbb{Z}_n$-manifold carries a well-defined fundamental class in $H_m(V;\mathbb{Z}_n)$. It is the nicest geometric model of a $\mathbb{Z}_n$-homology class.
(iii) A closed oriented manifold is a ($\mathbb{Z}_0$ or $\mathbb{Z}$)-manifold.

We return to the homotopy equivalence $f : (L, \partial L) \longrightarrow (M, \partial M)$. Let $g : V \longrightarrow M$ be a connected singular $\mathbb{Z}_n$-manifold in the interior of $M$, of dimension $v$. The graph of $g$ defines $V$ as a $\mathbb{Z}_n$-submanifold of the $\mathbb{Z}_n$-manifold $M \times V$.[1] Consider $\overline{f} = f \times (\text{identity on } V)$ mapping $(L, \partial L) \times V$ to $(M, \partial M) \times V$. If $\pi_1(M) = \pi_1(V) = \pi_1(\delta V) = 0$, $v = 2s$, and $\dim(M) \geq 3$, then we may deform $\overline{f}$ so that it has the following properties:

(i) $\overline{f}$ is transverse regular to $(V, \delta V) \subseteq M \times (V, \delta V)$ with $(U, \delta U) \subseteq L \times (V, \delta V)$ where $U = f^{-1}V$.

(*) (ii) $\overline{f} : \delta U \longrightarrow \delta V$ is a homotopy equivalence.[2]

(iii) $\overline{f} : U \longrightarrow V$ is $s$-connected where $v = \dim(V) = 2s$. See [S1] and [W1].

---

[1] Using the graph of $g$ is unnecessary if $g$ is an embedding. Note that this construction is the Gysin homomorphism for bordism.

[2] We assume further that $\overline{f} : \delta U \longrightarrow \delta V$ is a $PL$-homeomorphism in case $\dim(\delta U) = 3$. This is possible.

Let $K_s = \ker \overline{f}_* \subseteq H_s(U;\mathbb{Z})$. If $s$ is even $K_s$ admits a symmetric quadratic form (the intersection pairing) which is even ($\langle x,x\rangle$ is even) and non-singular. Thus $K_s$ has an index which is divisible by 8. If $s$ is odd, then $K_s \otimes \mathbb{Z}_2$ admits a symmetric quadratic form which has an Arf-Kervaire invariant in $\mathbb{Z}_2$.

We define the **splitting obstruction** of $f : (L, \partial L) \longrightarrow (M, \partial M)$ **along** $V$ by

$$\mathcal{O}_f(V) = \begin{cases} \text{Arf-Kervaire } (K_s) & \in \mathbb{Z}_2 \text{ if } s = 2k+1 \\ \frac{1}{8}\text{Index}(K_s) \text{ (modulo } n) & \in \mathbb{Z}_n \text{ if } s = 2k > 2 \\ \frac{1}{8}\text{Index}(K_s) \text{ (modulo } 2n) & \in \mathbb{Z}_{2n} \text{ if } s = 2. \end{cases}$$

We claim that $\mathcal{O}_f(V)$ only depends on the homotopy class of $f$. Also for $s \neq 2$, $\mathcal{O}_f(V) = 0$ iff $f$ may be deformed to a map split along $V$, i.e. $f^{-1}(V, \delta V)$ is homotopy equivalent to $(V, \delta V)$.

More generally we make the following:

**Definition 3.** Let $f : (L, \partial L) \longrightarrow (M, \partial M)$ be a homotopy equivalence and let $g : V \longrightarrow M$ be a singular variety in $M$. **The splitting invariant of $f$ along the variety** $V$ is the function which assigns to each component of $V$ the splitting obstruction of $f$ along that component.

Now we replace the Eilenberg obstruction theory of Theorem 1 by a first-order theory. We assume for simplicity that $Q$ is $\emptyset$.

**Theorem 2. (The Characteristic Variety Theorem)**
*Let $f : (L, \partial L) \longrightarrow (M, \partial M)$ be a homotopy equivalence as in Theorem 1. Then there is a (characteristic) singular variety in $M$, $V \longrightarrow M$, with the property that $f$ is homotopic to a piecewise linear homeomorphism iff the splitting invariant of $f$ along $V$ is identically zero.*

For example:

(i) (characteristic variety of $\mathbb{QP}^n$) $= (\mathbb{QP}^1 \cup \mathbb{QP}^2 \cup \ldots \cup \mathbb{QP}^{n-1} \xrightarrow{\text{inclusion}} \mathbb{QP}^n)$

(ii) (characteristic variety of $S^p \times S^q \times S^r$)

$=$ even dimensional components of

$(S^p \cup S^q \cup S^r \cup S^p \times S^q \cup S^p \times S^r \cup S^q \times S^r \xrightarrow{\text{inclusion}} S^p \times S^q \times S^r)$

(iii) (characteristic variety of $\mathbb{CP}^n$) $= (\mathbb{CP}^2 \cup \mathbb{CP}^3 \cup \ldots \cup \mathbb{CP}^{n-1} \xrightarrow{\text{inclusion}} \mathbb{CP}^n)$

(iv) (characteristic variety of a regular neighborhood $M$ of $S^{4k-1} \cup_r e^{4k}$)

$$= (V^{4k} \xrightarrow{\text{degree 1}} M)$$

where $V^{4k}$ is the $\mathbb{Z}_r$-manifold obtained from $S^{4k}$ by removing the $r$ open disks and identifying the boundaries.

We remark that there is not in general a canonical characteristic variety for $M$. We will discuss below conditions that insure that a variety in $M$ is characteristic and what choices are available.

First we consider the natural question raised by Theorem 2 – what are the relations on the set of all splitting invariants of homotopy equivalences $f : (L, \partial L) \longrightarrow (M, \partial M)$?

One relation may be seen by example – if $f : (L, \partial L) \longrightarrow (M, \partial M)$ is a homotopy equivalence and $S^4 \subseteq M$, then Index $f^{-1}(S^4) \equiv 0 \pmod{16}$ by a theorem of Rochlin. Thus the splitting obstruction of $f$ along $S^4$ is always even.

More generally, if $V$ is a singular variety in $M$, then a **four** dimensional component $N$ of $V$ is called a **spin component of $V$ in $M$** if:

(i) $N$ is a ($\mathbb{Z}$ or $\mathbb{Z}_{2^r}$)-manifold,

(ii) $\langle x \cup x, [N]_2 \rangle = 0$ for all $x \in H^2(M; \mathbb{Z}_2)$ where $[N]_2$ is the orientation class of $N$ taken mod 2.

Then we can state the following generalization of Theorem 2.

**Theorem 2′.** *Let* $(M, \partial M)$ *be a simply connected piecewise linear manifold pair with* $\dim(M) \geq 6$. *Then there is a* **characteristic singular variety $V$ in $M$** *with the following properties:*

(i) *Let* $g_i : (L_i, \partial L_i) \longrightarrow (M, \partial M)$ *be homotopy equivalences* $i = 0, 1$. *Then there is a piecewise linear homeomorphism* $c : L_0 \longrightarrow L_1$ *such that*

$$\begin{array}{ccc} (L_0, \partial L_0) & \xrightarrow{\ g_0\ } & \\ \cong \big\downarrow c & & (M, \partial M) \\ (L_1, \partial L_1) & \xrightarrow{\ g_1\ } & \end{array}$$

*is homotopy commutative iff*

$$(\textit{Splitting invariant of } g_0 \textit{ along } V) = (\textit{Splitting invariant of } g_1 \textit{ along } V)\,.$$

(ii) *A function on the components of $V$ with the proper range is the splitting invariant of a homotopy equivalence iff its values on the four dimensional spin components are even.*

Note that Theorem 2 follows from Theorem 2′ (i) by taking $g_0 = f$ and $g_1 =$ identity map of $(M, \partial M)$

$$\begin{array}{ccc} (L,\partial L) & \xrightarrow{\quad f \quad} & \\ \cong \downarrow c & & (M,\partial M) \\ (M,\partial M) & \xrightarrow{\text{identity}} & \end{array}$$

**Proof of Theorem 2**:

**The Kervaire Obstruction in $H^{4*+2}(M;\mathbb{Z}_2)$**

There is a very nice geometrical argument proving one half of the characteristic variety theorem. Namely, assume the homotopy equivalence $f : (L, \partial L) \longrightarrow (M, \partial M)$ can be deformed to a $PL$-homeomorphism on some neighborhood $Q$ of the $(k-1)$-skeleton of $L$ and $f(L - Q) \subseteq M - f(Q)$:

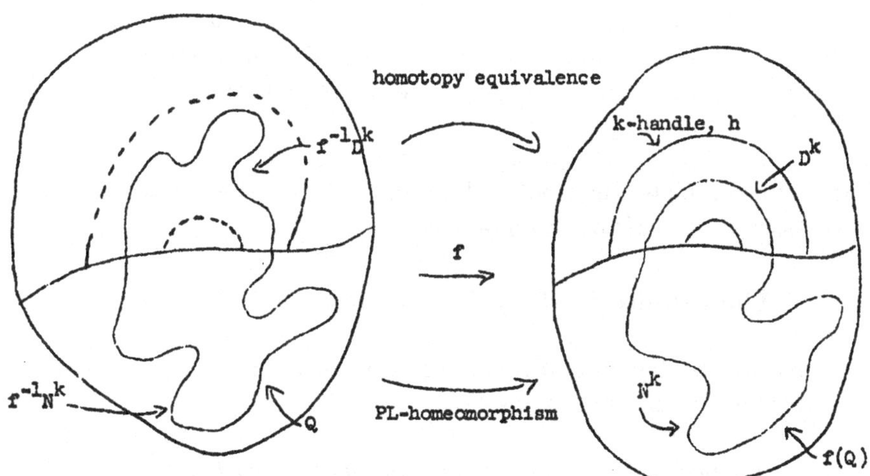


Figure 2

Suppose that $k = 4s + 2$ and recall that the obstruction class in

$$H^{4s+2}(M;\mathbb{Z}_2)$$

is represented by a cochain $c$ calculated by looking at various $(f^{-1}D^k)$'s – where the

$D^k$'s are the core disks of handles attached along $\partial f(Q)$. In fact, for a particular handle $h$, $c(D^k)$ is the Kervaire invariant of the framed manifold $f^{-1}(D^k)$ (which equals the class of $f^{-1}(D^k)$ in $P_{4s+2} = \mathbb{Z}_2$). See Figure 2.

*NOW* assume that there is a $\mathbb{Z}_2$-manifold $N^k$ embedded in $f(Q)$ union the $k$-handle $h$ which intersects the $k$-handle in precisely $D^k$. Then $f^{-1}N^k$ consists of two pieces – one is $PL$ homeomorphic to $N^k$ intersect $f(Q)$ and one is just $f^{-1}D^k$. Thus it is clear that the obstruction to deforming $f$ on all of $M$ so that $f^{-1}N^k$ is homotopy equivalent to $N^k$ is precisely $c(D^k) =$ Kervaire Invariant of $f^{-1}D^k$.

This means that $c(D^k)$ is *determined by the splitting obstruction of $f$ along $N^k$ - it does not depend on the deformation of $f$ to a $PL$-homeomorphism on $Q$.*

Roughly speaking, the part of $N^k$ in $f(Q)$ binds all possible deformations of $f$ together.

From cobordism theory [CF1] we know that any homology class in $H_k(M; \mathbb{Z}_2)$ is represented by a possibly singular $\mathbb{Z}_2$-manifold $N^k$ in $M$. So for part of the characteristic variety[1] we choose a collection of singular $\mathbb{Z}_2$-manifolds in $M$ of dimension $4s+2$, $2 \leq 4s+2 \leq \dim M$. We suppose that these represent a basis of $H_{4s+2}(M; \mathbb{Z}_2)$, $2 \leq 4s+2 \leq \dim M$.

The splitting obstructions for $f$ along these $\mathbb{Z}_2$-manifolds in $M$ determine homomorphisms $H_{4s+2}(M; \mathbb{Z}_2) \longrightarrow \mathbb{Z}_2$ which in turn determine cohomology classes in $H^{4s+2}(M; \mathbb{Z}_2)$. The argument above (generalized slightly)[2] shows that the lowest dimensional non-zero class among these is the first non-vanishing Eilenberg obstruction in dimension $4s+2$, (if it exists).

This would complete the proof of Theorem 2 if we did not have to cope with the obstructions in $H^{4i}(M; \mathbb{Z})$. So now the fun begins.

## The Infinite (Index) Obstructions in $H^{4*}(M; \mathbb{Z})$

Of course we can try to apply the argument of Figure 2 to characterize the Eilenberg obstructions in dimensions $4s$.

The attempt succeeds in *characterizing the Eilenberg obstructions in $H^{4s}(M; \mathbb{Z})$ modulo odd torsion elements.*

---

[1] We shall see below that some of the two dimensional components are not needed and others are replaced by certain four-dimensional (non-spin) components, e.g. $M = \mathbb{CP}^n$.

[2] The fact that the submanifolds are singular presents no difficulty – for we may look at graphs or cross the problem with a high dimensional disk.

Let $D^k$, $k = 4s$ be the core disk of a handle $h$ attached along the boundary of $f(Q)$ which determines an element of infinite order in $H^{4s}(M;\mathbb{Z})$. Then by obstruction theory in $MSO$ there is an oriented submanifold $N^k$ of $f(Q) \cup h$ which intersects the handle in a certain positive number of oriented copies of $D^k$. We can then calculate $[\frac{1}{8}$ Index of $f^{-1}D^k] =$ [class of $f^{-1}D^k$ in $P_{4s} = \mathbb{Z}]$[1] in terms of Index $f^{-1}N^k$ – Index $N^k$. The latter integer is determined by the splitting obstruction of $f$ along $N^k$. This characterizes the Eilenberg obstructions modulo torsion elements in $H^{4s}(M;\mathbb{Z})$.

## The 2-Torsion (Index) Obstructions in $H^{4*}(M;\mathbb{Z})$

Now suppose $D^k$ represents a generator of order $n$ in $H^{4s}(M;\mathbb{Z})$ $(k = 4s)$ and there is a singular $\mathbb{Z}_n$-manifold $N^k$ in $f(Q) \cup h$ which intersects the handle $h$ in $D^k$. Then the argument of Figure 2 again shows that the value of an Eilenberg obstruction cochain on $[D^k]$ taken mod $n$ is just the splitting invariant of $f$ along $N^k$.

From cobordism theory we can show that such an $N^k$ exists if $n$ is a power of 2.

So now we can characterize the Eilenberg obstructions modulo odd torsion elements. We add to our characteristic variety the manifolds considered in the previous two paragraphs – namely:

(i) an appropriate (as above) closed oriented manifold of dimension $4s$ for each element of a basis of $H^{4s}(M;\mathbb{Z})/\text{Torsion}$.[2]

(ii) an appropriate $\mathbb{Z}_{2^r}$-manifold of dimension $4s$ for each $\mathbb{Z}_{2^r}$-summand in $H^{4s}(M;\mathbb{Z})$.

All this for $4 \leq 4s < \dim M$.

The above applications of cobordism theory are based on the fact that the Thom spectrum for the special orthogonal group, $MSO$, has only *finite* $k$-invariants of *odd* order. (See [CF1]).

We also use the fact that the homotopy theoretical bordism homology with $\mathbb{Z}_n$-coefficients is just the geometric bordism homology theory defined by $\mathbb{Z}_n$-manifolds.

---

[1] Except when $4s = k = 4$, in which case we calculate $\frac{1}{16}$ Index $f^{-1}D^k$.
[2] We will later impose an additional restriction on these manifolds so that Theorem 2′ (ii) will hold.

Now the proof of Theorem 2 would be complete if $H^{4*}(M;\mathbb{Z})$ had no odd torsion.

## The Odd Torsion Obstructions in $H^{4*}(M;\mathbb{Z})$, Manifolds With Singularities, and $k$-Homology

We have reduced our analysis to the case when the Eilenberg obstructions are concentrated in the odd torsion subgroup of $H^{4*}(M;\mathbb{Z})$.

However, we are stopped at this point by the crucial fact that $\mathbb{Z}_n$-manifolds are not general enough to represent $\mathbb{Z}_n$-homology when $n$ is odd. (For example the generator of $H_8(K(\mathbb{Z},3);\mathbb{Z}_3)$ is not representable.) Thus the crucial geometrical ingredient of the "Figure 2" proof is missing.

In the $n$ odd case we can change the format of the proof slightly. Let $D^k$ be the core disk of a $k$-handle $h$ representing a generator of *odd* order $n$ in $H^{4s}(M;\mathbb{Z})$. Let $N^k$ be a $\mathbb{Z}_n$-manifold in $f(Q)\cup h$ situated as usual. *Then we claim that the class of $f^{-1}D^k$ in $P_k\otimes\mathbb{Z}_n$ is determined by* Index $f^{-1}N^k$ – Index $N^k\in\mathbb{Z}_n$.[1] (We can recover the $\frac{1}{8}$ factor since $n$ is odd.) But the index of $f^{-1}N^k$ only depends on the homotopy class of $f$ because of transversality and the cobordism invariance of the mod $n$ index. ([N2])

Thus we see that the rigidity of the odd torsion Eilenberg obstruction "follows" from the existence of a **geometrical** $\mathbb{Z}_n$-manifold object which:

(i) is general enough to represent $\mathbb{Z}_n$-homology,

(ii) is nice enough to apply transversality,

(iii) has an additive index $\in\mathbb{Z}_n$ which is a cobordism invariant and which generalizes the usual index.

Finding a reasonable solution of (i) is itself an interesting problem.*

We proceed as follows. Let $\mathbb{C}_1,\mathbb{C}_2,\ldots$ denote a set of ring generators for smooth bordism modulo torsion, $\Omega_*/\text{Torsion}$ ($\dim\mathbb{C}_i=4i$). We say that a polyhedron is "like" $S^n*\mathbb{C}_1$ ($=S^n$ join $\mathbb{C}_1$) if it is of the form $W\cup L\times\text{cone}\,\mathbb{C}_1$, $W$ a $PL$-manifold, $\partial W=L\times\mathbb{C}_1$, i.e. has a singularity structure like $S^n*\mathbb{C}_1$. More generally we say that a polyhedron is "like" $Q=S^n*\mathbb{C}_{i_1}*\mathbb{C}_{i_2}*\ldots*\mathbb{C}_{i_r}$[2] if it admits a global decomposition "like" $Q$.

---

[1] The index of a $\mathbb{Z}_n$-manifold $(N^{4k},\delta N)$ is the index of $N/\delta N$ taken modulo $n$. The index of $N/\delta N$ is the signature of the (possibly degenerate) cup product pairing on $H^{2k}(N/\delta N;\mathbb{Q})$.

* A solution is given by Rourke, Bull. L.M.S. **5** (1973) 257–262

[2] We require $i_1<i_2<\ldots<i_r$.

For example, $S^n * \mathbb{C}_1 * \mathbb{C}_2$ may be decomposed as in Figure 3.

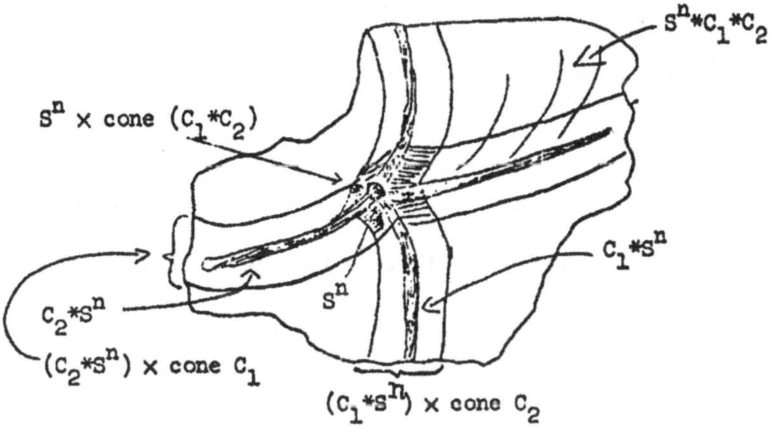


Figure 3

These polyhedra play the role of closed "*manifolds*" in our theory. Part of their structure is the join-like decomposition of a neighborhood of the singularity set together with a compatible linear structure on the stable tangent bundle of the complement of the singular set.

"Manifolds with boundary" and "$\mathbb{Z}_n$-manifolds" are easy generalizations.

(i) *The bordism homology theory defined by these manifolds with singularities is usual integral homology theory.*
(ii) Also transversality and other geometrical constructions are fairly easy with these varieties.
(iii) They do *not however have a good index.* Our mistake came when we introduced the cone on $\mathbb{C}_1$ = cone on $\mathbb{CP}^2$, say.

If we make the analogous construction using only $\mathbb{C}_2, \mathbb{C}_3, \mathbb{C}_4, \ldots$ where Index $\mathbb{C}_i = 0$ $i = 2, 3, 4, \ldots$ then *we can define a proper index.*

However, *we no longer have ordinary homology theory but a theory $V_*$ such that $V_*(\text{pt.})$ is a polynomial* algebra on one 4-dimensional generator $[\mathbb{CP}^2]$. $V_*$ is in fact a geometric representation of connective $k$-homology and the natural transformation $\Omega_* \longrightarrow V_*$[1] is closely related to the transformation $I : \Omega_* \longrightarrow K_*$ constructed below.[2]

[1] This is obtained by regarding a non-singular manifold as a variety.
[2] We are working modulo 2-torsion in this paragraph.

Thus we see that the Eilenberg obstructions in dimension $4k$ are not well-defined. Their values may be varied on those $\mathbb{Z}_n$-classes whose geometric representative requires a $\mathbb{CP}^2$-singularity (i.e. does not come from $V_*$). This may be seen quite clearly in dimension 8. In fact, from the homotopy theory below we see that the $H^{4k}(M;\mathbb{Z}_n)$ modulo the indeterminacy of the Eilenberg obstructions (reduced mod $n$) is precisely dual to the subgroup of $V_*$ representable elements in $H_{4k}(M;\mathbb{Z}_n)$.

This duality may also be seen geometrically but it is more complicated.

## II. The Characteristic Bundle of a Homotopy Equivalence

The proof of Theorem 2 (the Characteristic Variety Theorem) can be completed by studying the obstruction theory of Theorem 1 from the homotopy theoretical point of view.

**Definition**. ($F/PL$-bundle, $F/O$-bundle). An **$F/PL)_n$-bundle** over a finite complex $X$ is a (proper) homotopy equivalence $\theta : E \longrightarrow X \times \mathbb{R}^n$ where $\pi : E \longrightarrow X$ is a piecewise linear $\mathbb{R}^n$-bundle and

$$\begin{array}{ccc} E & \xrightarrow{\ \theta\ } & X \times \mathbb{R}^n \\ {\scriptstyle \pi}\big\downarrow & & \big\downarrow{\scriptstyle p_1} \\ X & \xrightarrow{\ \text{identity}\ } & X \end{array}$$

is homotopy commutative.

Two $F/PL)_n$-bundles $\theta_0$ and $\theta_1$ are **equivalent** iff there is a piecewise linear bundle equivalence $b : E_0 \longrightarrow E_1$ so that

$$\begin{array}{ccc} E_0 & \searrow^{\theta_0} & \\ {\scriptstyle b}\big\downarrow & & X \times \mathbb{R}^n \\ E_1 & \nearrow_{\theta_1} & \end{array}$$

is properly homotopy commutative.

An $F/O)_n$-bundle is the corresponding linear notion.[1]

[1] These bundle theories are classified [B1] by the homotopy classes of maps into certain $CW$ complexes $F/PL)_n$ and $F/O)_n$. The correspondence $\theta \longrightarrow \theta \times \text{identity}_{\mathbb{R}}$ defines stabilization maps $F/PL)_n \longrightarrow F/PL)_{n+1}$, $F/O)_n \longrightarrow F/O)_{n+1}$. The stable limits are denoted by $F/PL$ and $F/O$ respectively. Using a "Whitney sum" operation $F/O$ and $F/PL$ become homotopy associative, homotopy commutative $H$-spaces.

Let $g : (L, \partial L) \longrightarrow (M, \partial M)$ be a homotopy equivalence of compact piecewise linear manifolds with homotopy inverse $\overline{g} : (M, \partial M) \longrightarrow (L, \partial L)$.

A **characteristic** $F/PL)_n$**-bundle of** $g$ is any composition $\theta_g$, given by

$$E \xrightarrow[\cong]{c\, = PL\text{-homeomorphism}} L \times \mathbb{R}^n \xrightarrow{g \times \text{identity}} M \times \mathbb{R}^n ,$$

where $E$ is the normal bundle of an embedding $M \subset L \times \mathbb{R}^n$ ($n \gg \dim L$) approximating $\overline{g} \times 0$. $c$ is any identification (homotopic to $\overline{g}$) of the total space of $E$ with all of $L \times \mathbb{R}^n$. ($c$ may be constructed for example á lá Mazur using the "half-open" $h$-cobordism theorem.)

Notice that $\theta_g$ is transverse regular to $M \times 0$ with inverse image $PL$-homeomorphic to $L$

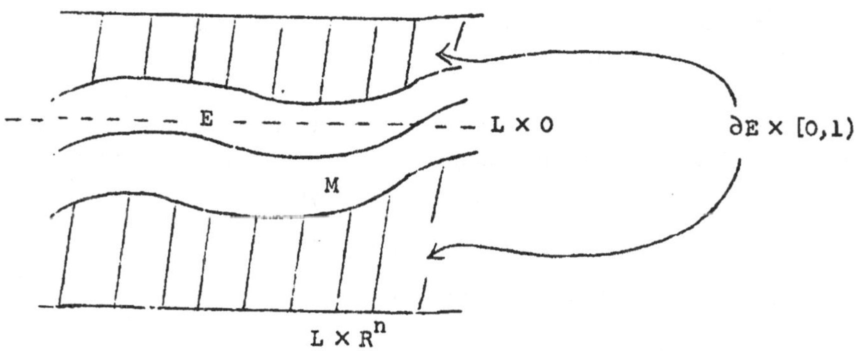


**The characteristic bundle of the homotopy equivalence** $g : (L, \partial L) \longrightarrow (M, \partial M)$ is the stable equivalence class of $\theta_g$ considered as a homotopy class of maps

$$\theta_g \ : \ M \longrightarrow F/PL \ .$$

If $L$ and $M$ are smooth, $E$ is a vector bundle, $c$ will be a diffeomorphism, and we obtain the characteristic $F/O$-bundle of a homotopy equivalence between smooth manifolds

$$\eta_g \ : \ M \longrightarrow F/O \ .$$

## The Classification of $h$-Triangulations and $h$-Smoothings

To state the homotopy theoretical analogue of Theorem 1 we consider a homotopy equivalence of a $PL$-manifold pair with $(M, \partial M)$ as defining a "**homotopy-triangulation**" of $(M, \partial M)$. Two **$h$-triangulations** $g_0 : (L_0, \partial L_0) \longrightarrow (M, \partial M)$ and $g_1 : (L_1, \partial L_1) \longrightarrow (M, \partial M)$ are "concordant" iff there is a $PL$-homeomorphism

$c : (L_0, \partial L_0) \longrightarrow (L_1, \partial L_1)$ so that

$$\begin{array}{ccc} (L_0, \partial L_0) & \xrightarrow{g_0} & \\ \cong \downarrow c & & (M, \partial M) \\ (L_1, \partial L_1) & \xrightarrow{g_1} & \end{array}$$

is homotopy commutative. We denote the set of concordance classes of $h$-triangulations of $M$ by $hT(M)$.

Note that the characteristic variety theorem asserts that the concordance class of an $h$-triangulation $g : (L, \partial L) \longrightarrow (M, \partial M)$ is *completely* determined by the splitting invariant of $g$.

In a similar fashion we obtain the set of concordance classes of **$h$-smoothings** of $M$, $hS(M)$.

The zero element in $hT(M)$ or $hS(M)$ is the class of id. : $M \longrightarrow M$. The characteristic bundle construction for a homotopy equivalence defines transformations

$$\theta : hT(M) \longrightarrow (M, F/PL)$$
$$\eta : hS(M) \longrightarrow (M, F/O)$$

where $(X, Y)$ means the set of homotopy classes of maps from $X$ to $Y$.

Assume $\pi_1(M) = \pi_1(\partial M) = 0$, $n = \dim M \geq 6$.

**Theorem 3.** *If $\partial M \neq \emptyset$, then*

$$\theta : hT(M) = \left\{ \begin{array}{c} \textit{concordance classes of} \\ h\textit{-triangulations of } M \end{array} \right\} \longrightarrow (M, F/PL)$$

*and*

$$\eta : hS(M) = \left\{ \begin{array}{c} \textit{concordance classes of} \\ h\textit{-smoothings of } M \end{array} \right\} \longrightarrow (M, F/O)$$

*are isomorphisms.*
*If $\partial M = \emptyset$, we have the exact sequences (of based sets)*

$$\text{(i)}\quad 0 \longrightarrow hT(M) \xrightarrow{\theta} (M, F/PL) \xrightarrow{\mathcal{S}} P_n$$

$$\text{(ii)}\quad \theta_n \partial\pi \xrightarrow{\#} hS(M) \xrightarrow{\eta} (M, F/O) \xrightarrow{\mathcal{S}} P_n \,.$$

**Proof.** See [S1].

Here $\mathcal{S}$ is the surgery obstruction for an $F/PL$ or $F/O$ bundle over a closed (even dimensional) manifold; and $\#$ is obtained from the action of $\theta_n$ on $hS(M)$,

$(g : L \longrightarrow M) \mapsto (g : L\#\Sigma \longrightarrow M)$.

$\mathcal{S}$ will be discussed in more detail below. We remark that the exactness of (ii) at $hS(M)$ is stronger, namely

$$\{\text{orbits of } \theta_n \partial\pi\} \;\cong\; \text{image}\, \eta \;.$$

Also (i) may be used to show that $\partial M = \emptyset$ implies the:

**Corollary**. *If $M$ is closed then $\theta : hT(M) \cong (M - \text{pt.}, F/PL)$.*

Easy transversality arguments show that

$$\pi_i(F/PL) \;=\; P_i \;\;,\;\; \pi_i(F/O) \;=\; A_i \;.$$

Thus the Theorem 1 obstructions in

$$H^i(M; A_i) \;\text{ or }\; H^i(M; P_i)$$

for deforming $g : (L, \partial L) \longrightarrow (M, \partial M)$ to a diffeomorphism or a $PL$-homeomorphism become the homotopy theoretical obstructions in

$$H^i(M; \pi_i(F/O)) \;\text{ or }\; H^i(M; \pi_i(F/PL))$$

for deforming $\eta_g$ or $\theta_g$ to the point map.

In fact using naturality properties of $\theta$ and $\eta$ we can precisely recover the obstruction theory of Theorem 1 from statements about the "kernels" of $\theta$ and $\eta$ given in Theorem 3.

We obtain *new* information from the statements about the "cokernels" of $\theta$ and $\eta$ given in Theorem 3. For example, if we consider the map

$$\mathbb{CP}^4 - \text{pt.} \;\cong\; \mathbb{CP}^3 \xrightarrow{\text{deg } 1} S^6 \xrightarrow{\text{gen } \pi_6} F/PL$$

we obtain an interesting $h$-triangulation of $\mathbb{CP}^4$, $M^8 \longrightarrow \mathbb{CP}^4$.

Now we may study the obstruction theory of Theorem 1 by studying the homotopy theory of $F/O$ and $F/PL$.

For example using the fact that $F/O$ and $F/PL$ are $H$-spaces (Whitney sum) (and thus have trivial $k$-invariants over the rationals) one sees immediately that the triangulating and smoothing obstructions for a homotopy equivalence $f$ are torsion cohomology classes iff $f$ is a correspondence of rational Pontrjagin classes.

To describe the obstructions completely we must again restrict to the piecewise linear case.

## The Homotopy Theory of $F/PL$

We have already seen that the homotopy groups of $F/PL$ are very nice:

| $i$ | 1 | 2 | 3 | 4 | 5 | 6 | 7 | 8 | ... |
|---|---|---|---|---|---|---|---|---|---|
| $\pi_i(F/PL)$ | 0 | $\mathbb{Z}_2$ | 0 | $\mathbb{Z}$ | 0 | $\mathbb{Z}_2$ | 0 | $\mathbb{Z}$ | ... |

This regularity is also found in the global homotopy structure of the space.

For describing this structure we will localize $F/PL$ *at* the prime 2 and then *away* from the prime 2.

If $X$ is a homotopy associative homotopy commutative $H$-space, then

"$X$ localized at 2" $\equiv X_{(2)}$ is the $H$-space which represents the functor

$$(\ ,X)\otimes\mathbb{Z}_{(2)}$$

where $\mathbb{Z}_{(2)} \ = \ \mathbb{Z}[\frac{1}{3},\frac{1}{5},\ldots,\frac{1}{p_i},\ldots]$, $p_i$ the $i^{\text{th}}$ odd prime.

"$X$ localized away from 2" $\equiv X_{(\text{odd})}$ is the $H$-space which represents the functor

$$(\ ,X)\otimes\mathbb{Z}_{(\text{odd})}$$

where $\mathbb{Z}_{(\text{odd})} \ = \ \mathbb{Z}[\frac{1}{2}]$.

Note that there are natural projections $p_{(2)}$ and $p_{(\text{odd})}$

$$\begin{array}{ccccc} & & & & X_{(2)} \\ & & & \overset{p_{(2)}}{\nearrow} & \\ Y = CW\text{-complex} & \xrightarrow{\ f\ } & X & & \\ & & & \underset{p_{(\text{odd})}}{\searrow} & \\ & & & & X_{(\text{odd})} \end{array}$$

Also $f : Y \longrightarrow X$ is homotopic to zero iff $p_{(2)} \circ f$ and $p_{(\text{odd})} \circ f$ are homotopic to zero. Thus it suffices to study $X_{(2)}$ and $X_{(\text{odd})}$.

Let $BO$ denote the classifying space for stable equivalence classes of vector bundles over finite complexes.

Let $K(\pi, n)$ denote the Eilenberg-MacLane space, having one non-zero homotopy group $\pi$ in dimension $n$.

Let $\delta Sq^2$ denote the unique element of order 2 in $H^5(K(\mathbb{Z}_2,2);\mathbb{Z}_{(2)})$, and $K(\mathbb{Z}_2,2)\times_{\delta Sq^2} K(\mathbb{Z}_{(2)},4)$ the total space of the principal fibration over $K(\mathbb{Z}_2,2)$ with $K(\mathbb{Z}_{(2)},4)$ as fibre and principal obstruction ($k$-invariant) $\delta Sq^2$.

Then we have the following:

**Theorem 4.** (i) *$F/PL_{(2)}$ is homotopy equivalent to*

$$\prod \; = \; K(\mathbb{Z}_2,2)\times_{\delta Sq^2} K(\mathbb{Z}_{(2)},4)\times\prod_{i=1}^{\infty} K(\mathbb{Z}_2,4i+2)\times K(\mathbb{Z}_{(2)},4i+4)\ .$$

(ii) *$F/PL_{(\mathrm{odd})}$ is homotopy equivalent to $BO_{(\mathrm{odd})}$.*
(iii) *$F/PL$ is homotopy equivalent to the fibre product of $i$ and $ph$ in the diagram*

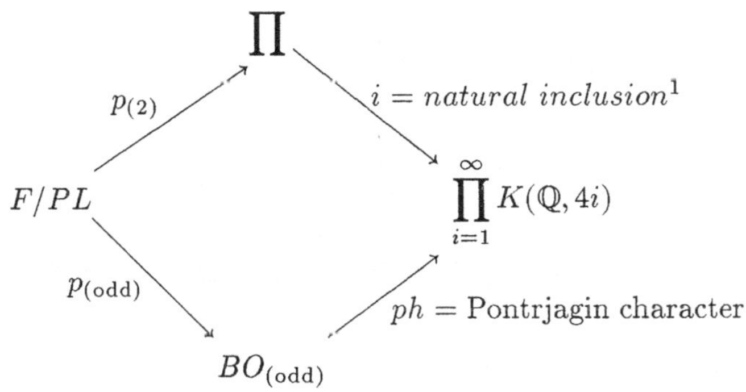


**Corollary**. *$H^*(F/PL;\mathbb{Z})$ has no odd torsion and the 2-torsion may be calculated.*

**Corollary**. *$\mathbb{Z}\times F/PL$ is an infinite loop space. In fact it is homotopy equivalent to the 0th space in the $\Omega$-spectrum of a multiplicative cohomology theory.*

**Corollary**. *If $\mathbb{O}=\Omega(F/PL)=$ loop space of $F/PL$, then $\mathbb{O}$ satisfies a Bott periodicity of length four, namely*

$$\Omega^4\mathbb{O}\;\cong\;\mathbb{O}\,,\quad \textit{as } H\textit{-spaces}\ .$$

We use the notation $\mathbb{O}$ because $\mathbb{O}_{(\mathrm{odd})}\cong O_{(\mathrm{odd})}$ where $O$ is the infinite orthogonal group.

## Bordism, Homology Theory, and $K$-Theory

In order to prove and apply Theorem 4 we need to study the relationship between smooth bordism and ordinary homology on the one hand (for the prime 2) and smooth bordism and $K$-theory on the other hand (for odd primes). Recall that $\Omega_*(X)$ is a module over $\Omega_*=\Omega_*(\mathrm{pt})$ by the operation

$$(f:M\longrightarrow X,N)\longrightarrow(fp_2:N\times M\longrightarrow X)\ .$$

[1] On $\pi_4$ $i_*$ is twice the natural embedding $\mathbb{Z}_{(2)}\longrightarrow\mathbb{Q}$.

$\mathbb{Z}$ is a module over $\Omega_*$ by Index : $\Omega_* \longrightarrow \mathbb{Z}$.
Then we can form $\Omega_*(X) \otimes_{\Omega_*} \mathbb{Z}$, and obtain a $\mathbb{Z}_4$-graded *functor.*
Let $K_*(X)$ denote the $\mathbb{Z}_4$-graded homology theory defined by $KO_*(X) \otimes \mathbb{Z}_{(\text{odd})}$.
$KO_*$ is the homology theory dual to real $K$-theory, $KO^*(X)$.

**Theorem 5.** *There are natural equivalences of functors:*

(i) $\Omega_*(X) \otimes \mathbb{Z}_{(2)} \cong H_*(X; \Omega_* \otimes \mathbb{Z}_{(2)})$ *as $\mathbb{Z}$-graded $\Omega_*$-modules,*

(ii) $\Omega_*(X) \otimes_{\Omega_*} \mathbb{Z}_{(\text{odd})} \cong KO_*(X) \otimes \mathbb{Z}_{(\text{odd})} \equiv K_*(X)$ *as $\mathbb{Z}_4$-graded $\mathbb{Z}$-modules.*

**Proof.** (i) can be found in [CF1]. (ii) is analogous to [CF2] but different (and simpler) because the module structure on the left is different. ([CF2] uses the $\widehat{A}$ genus and the first cobordism Pontrjagin class.)

**Proof of (ii):** We construct a (multiplicative) transformation

$$I : \Omega_*(X) \longrightarrow K_*(X)$$

which on the point is essentially the index.

$I$ was first constructed by introducing the singularities described above into cobordism theory and then taking a direct limit.

It can also be constructed by first producing an element in $K^0(BSO)$ whose Pontrjagin character is

$$\frac{\widehat{A}}{L} = \frac{\widehat{A}\text{-genus}}{\text{Hirzebruch } L\text{-genus}}$$

(this is a calculation), and then applying the usual Thom isomorphism to obtain the correct element in $K^0(MSO)$.

$I$ induces a transformation

$$I : \Omega^0(X) \longrightarrow K^0(X)$$

which is in turn induced by a map of universal spaces

$$I : \Omega^\infty MSO_{(\text{odd})} \longrightarrow \mathbb{Z} \times BO_{(\text{odd})} .$$

$I$ is onto in homotopy (since there are manifolds of index 1 in each dim $4k$), thus the fibre of $I$ only has homotopy in dimensions $4k$. Obstruction theory implies that $I$ has a cross section. Therefore the transformation induced by $I$

$$I : \Omega^*(X) \otimes_{\Omega^*} \mathbb{Z}_{(\text{odd})} \longrightarrow K^*(X)$$

is onto for dimension zero cohomology.

When $X = MSO$, $I$ is easily seen to be an isomorphism. This fact together with the cross section above implies that $I$ is injective for dimension zero. (ii) now follows by Alexander duality and the suspension isomorphism.

The point of Theorem 5 is the following – the theories on the left have a nice geometrical significance for our problems while those on the right are nice algebraically.

Of course $H_*(X;\Omega_*)$ is constructed from a chain complex. We thus have the classical duality theorems (universal coefficient theorems) relating $H_*$ and $H^*$.

*These results are also true for $K_*$ and $K^*$.* For example, the multiplicative structure (see [A]) in $K^*$ defines homomorphisms

$$\begin{aligned} e\ &: K^0(X) \longrightarrow \mathrm{Hom}(K_0(X), \mathbb{Z}_{(\mathrm{odd})}) \\ &= \\ e_n\ &: K^0(X) \longrightarrow \mathrm{Hom}(K_0(X;\mathbb{Z}_n), \mathbb{Z}_n)\ ,\ n \text{ odd}\ . \end{aligned}$$

**Theorem 6.** *If $X$ is finite*
(i) *$e$ is onto,*
(ii) *if $\sigma \in K^0(X)$, then $\sigma = 0$ iff $e_n(\sigma) = 0$ for all odd integers $n$,*
(iii) *any compatible (w.r.t. $\mathbb{Z}_n \longrightarrow \mathbb{Z}_{n'}$) set of homomorphisms $(f, f_n)$ determines an element $\sigma$ in $K^0(X)$ such that $e_n(\sigma) = f_n$, $e(\sigma) = f$.*

(i) and (ii) were first proved by the author using intersection theory and the geometrical interpretation of $I$. (The hard part was to construct $e$ and $e_n$.)

However, using the multiplication in $K^0(X)$ coming from the tensor product of vector bundles (plus the extension to $\mathbb{Z}_n$-coefficients in [AT]) (i), (ii), and (iii) follow immediately from Bott periodicity and general nonsense.

The duality theorems for $K^*$ and $K_*$ were first proved by Anderson [A1]. We denote by $\Omega_*(X;\mathbb{Z}_n)$ the homology theory defined by bordism of $\mathbb{Z}_n$-manifolds. We make $P_* = 0, \mathbb{Z}_2, 0, \mathbb{Z}, \ldots$ into a *$\Omega_*$-module* by Index : $\Omega_* \longrightarrow P_*$. Then we have:

**Theorem 7. (The splitting obstruction of an $F/PL$ bundle)**
*There are* onto $\Omega_*$-module homomorphisms $\mathcal{S}$ *and* $\mathcal{S}_n$ *so that the following square commutes*

$$\begin{array}{ccc} \Omega_*(F/PL) & \xrightarrow{\ \ \mathcal{S}\ \ } & P_* \\ \text{natural inclusion}\ \Big\downarrow & & \Big\downarrow\ \text{reduction mod } n \\ \Omega_*(F/PL;\mathbb{Z}_n) & \xrightarrow{\ \ \mathcal{S}_n\ \ } & P_* \otimes \mathbb{Z}_n \end{array}$$

*The composition*

$$\pi_*(F/PL) \xrightarrow{\text{Hurewicz}} \Omega_*(F/PL) \xrightarrow{\ \mathcal{S}\ } P_*$$

*is an isomorphism if $* \neq 4$, and multiplication by 2 if $* = 4$.*

Using Theorem 5 (i) and $\mathcal{S}_2 : \Omega_{4*+2}(F/PL;\mathbb{Z}_2) \longrightarrow \mathbb{Z}_2$ which is more generally a $\Omega_*(\mathrm{pt};\mathbb{Z}_2)$-module homomorphism[1] we obtain:

**Corollary 1.** (**A Formula for the Kervaire Invariant of an $F/PL$-bundle over a $\mathbb{Z}_2$-manifold**). *There is a unique class*

$$\mathcal{K} \;=\; k_2 + k_6 + k_{10} + \ldots \in H^{4*+2}(F/PL;\mathbb{Z}_2)$$

*such that for any $\mathbb{Z}_2$-manifold in $F/PL$*

$$f \;:\; M^{4k+2} \longrightarrow F/PL$$

*we have*

$$\mathcal{S}_2(M^{4k+2}, f) \;=\; W(M)\cdot f^*\mathcal{K}[M]\ .$$

Using Theorem 5 (i) and $\mathcal{S} : \Omega_{4*}(F/PL) \longrightarrow \mathbb{Z}$ and working modulo torsion we obtain:

**Corollary 2.** *There is a class*

$$\mathcal{L} \;=\; \ell_4 + \ell_8 + \ell_{12} + \ldots \in H^{4*}(F/PL;\mathbb{Z}_{(2)})$$

*which is unique modulo torsion, such that*

$$\mathcal{S}(M^{4k}, f) \;=\; L(M)\cdot f^*\mathcal{L}[M]\ .$$

Here $W(M)$ and $L(M)$ are respectively the total Stiefel Whitney class and total Hirzebruch class.

The point of Corollary 2 is that $\mathcal{L}$ is a class with $\mathbb{Z}_{(2)}$-coefficients. $\mathcal{L}$ regarded as a class with rational coefficients is familiar, namely

$$\mathcal{L} \;=\; \ell_4 + \ell_8 + \ldots \;=\; \frac{1}{8} j^*(L_1 + L_2 + \ldots)$$

where $j : F/PL \longrightarrow B_{PL}$ is the natural map and $L_i$ is the universal Hirzebruch class in $H^{4i}(B_{PL};\mathbb{Q})$.

Now use

$$\begin{array}{ccc} & \Omega_*(F/PL) & \\ {\scriptstyle \text{natural projection}}\swarrow & & \searrow{\scriptstyle \mathcal{S}} \\ K_0(F/PL) & \xrightarrow{\ \mathcal{S}\otimes_{\Omega_*}\mathbb{Z}_{(\mathrm{odd})}\ } & \mathbb{Z}_{(\mathrm{odd})} \end{array}$$

(with $K_0(F/PL) \cong \Omega_{4*}(F/PL)\otimes_{\Omega_*}\mathbb{Z}_{(\mathrm{odd})}$) and Theorem 6 (i) to obtain:

[1] Via the mod 2 Euler characteristic $\Omega_*(\mathrm{pt},\mathbb{Z}_2) \longrightarrow \mathbb{Z}_2$.

**Corollary 3.** *There is a unique element $\sigma \in \widetilde{K}^0(F/PL)$ such that the Pontrjagin character of $\sigma = \mathcal{L}$ in $H^{4*}(F/PL; \mathbb{Q})$.*

Uniqueness follows from existence – which implies that $\widetilde{K}^0(F/PL)$ is a free $\mathbb{Z}_{(\mathrm{odd})}$-module.

**Proof of Theorem 7.** If $\theta : E \longrightarrow M \times \mathbb{R}^n$ is an $F/PL$-bundle over a $\mathbb{Z}_n$-manifold $M^{2s}$, then suppose $\theta^{-1}(M, \delta M)$ has the properties of $(U, \delta U)$ in $(*)$ (in the construction of the splitting obstruction in Section I) and define $\mathcal{S}(M, f)$ by

$$\mathcal{S}(M,f) = \begin{cases} \frac{1}{8}\{\mathrm{Index}(U,\delta U) - \mathrm{Index}(M,\delta M)\} \pmod n \\ \qquad s \text{ even}^1 \\ \text{Kervaire invariant of } \theta : (U,\delta U) \longrightarrow (M,\delta M) \\ \qquad s \text{ odd}. \end{cases}$$

**Remark.** The fact that $\mathcal{S} : \Omega_4(F/PL) \longrightarrow \mathbb{Z}$ is onto[2] while $\pi_4(F/PL) \longrightarrow \Omega_4(F/PL) \xrightarrow{\mathcal{S}} \mathbb{Z}$ is multiplication by 2 implies the first $k$-invariant of $F/PL$ in $H^5(K(\mathbb{Z}_2, 2); \mathbb{Z}) = \mathbb{Z}_4$ is *non-zero.*

Since $F/PL$ is an $H$-space the reduction of the $k$ invariant to $\mathbb{Z}_2$ coefficients must be primitive and have $Sq^1$ zero. It is therefore zero (by an easy calculation – pointed out to me by Milnor). This singles out $\delta Sq^2$ = "integral Bockstein of square two" as the first $k$-invariant of $F/PL$ (= first $k$-invariant of $BSO$, $F/O$, $BSPL$ etc.)

**Proof of Theorem 4:**

I. The $\sigma \in \widetilde{K}_0(F/PL)$ of Corollary 3 determines

$$p_{(\mathrm{odd})} : F/PL \longrightarrow BO_{(\mathrm{odd})} .$$

II. Using $\mathcal{K}$ and $\mathcal{L}$ of Corollaries 1 and 2 and the remark above we construct[3]

$$p_{(2)} : F/PL \longrightarrow \prod ,$$

---

[1] If $\dim M = 4$ $\mathcal{S}(M, f)$ is well-defined (modulo $2n$) if cobordisms of $\delta M$ are restricted to spin manifolds.

[2] 24 times the canonical complex line bundle over $\mathbb{CP}^2$ is fibre homotopically trivial.

[3] $i$ is onto in homotopy except in dimension 4 where it has index 2

so that

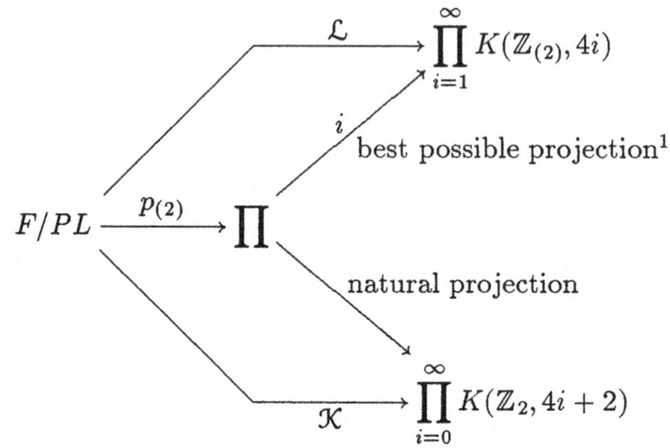


is homotopy commutative.

A calculation shows that $p_{(\mathrm{odd})}$ and $p_{(2)}$ are correct in homotopy. It is clear from the construction that

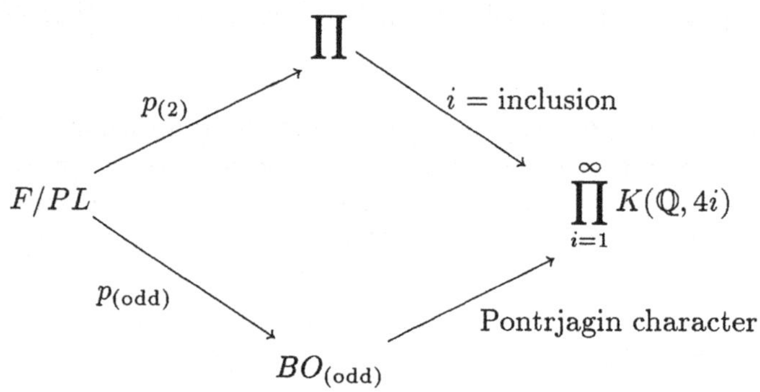


is homotopy commutative, so (iii) is proven.

**Remark.** The only part of the construction of the localizing projections $p_{(2)}$ and $p_{(\mathrm{odd})}$ which is not completely canonical is the construction of $\mathcal{L}$ used in the definition of $p_{(2)}$. $\mathcal{L}$ was only determined modulo torsion (this can be improved to "modulo torsion elements divisible by 2"). This difficulty arises from the lack of a nice geometrical description of the product of two $\mathbb{Z}_n$-manifolds (as a $\mathbb{Z}_n$-manifold).

We can make this aspect of $F/PL$-homotopy theory more intrinsic by formulating the results in terms of a characteristic variety.

If $X$ is a finite complex and $g : V \longrightarrow X$ is a singular variety in $X$, then for each $F/PL$-bundle over $X$ we can by restriction to $V$ associate a splitting obstruction on each component of $V$. We use the splitting invariant defined in the proof of Theorem 7 (with the refinement in dimension 4). We obtain a "function on $V$" for each $F/PL$-bundle over $X$.

Notice that no fundamental group hypothesis need be made to define the "splitting invariant along $V$" because every element in $\Omega_{2*}(F/PL;\mathbb{Z}_n)$ is represented by a simply connected pair $(M,\delta M)$.

**Theorem 4′. (The characteristic variety theorem for $F/PL$).**
*Let $X$ be a finite complex. Then there is a characteristic variety in $X$, $g : V \longrightarrow X$ with the property that:*

(i) *two $F/PL$-bundles over $X$ are equivalent iff their "splitting invariants along $V$" are equal.*

(ii) *a "function on $V$" is the splitting invariant of a bundle iff its values on the 4-dimensional* spin components[1] *of $V$ are even.*

**Remark.** It is easy to see that if $h : (M,\partial M)\longrightarrow(L,\partial L)$ is a homotopy equivalence then the splitting invariant of $h$ along a singular variety $V$ in $M$ is the same as the splitting invariant of $\theta_h$ along $V$. Thus Theorems 3 and 4′ imply Theorem 2′.

**Proof of Theorem 4′.** We first describe a suitable characteristic variety.

(i) Choose a collection of $(4i+2)$-dimensional $\mathbb{Z}_2$-manifolds in $X$, $f : \bigcup_i K_i \longrightarrow X$, so that $\{f_*(\text{fundamental class } K_i)\}$ is a basis of $A \bigoplus_{i>0} H_{4i+2}(X;\mathbb{Z}_2)$ where $A \subseteq H_2(X;\mathbb{Z}_2)$ is a subgroup dual to $\ker(Sq^2 : H^2(X;\mathbb{Z}_2)\longrightarrow H^4(X;\mathbb{Z}_2))$.

(ii) Choose a collection of $4i$-dimensional $\mathbb{Z}_{2^r}$-manifolds $f : \bigcup_j N_j \longrightarrow X$ such that $\{f_*(\delta N_j)\}$ is a basis of [2-torsion $\bigoplus_i H_{4i-1}(M;\mathbb{Z})$].

(iii) Choose a collection $\{V_\alpha\}$ of $4N$-dimensional $\mathbb{Z}_{p^r}$-manifolds for each odd prime $p$, $\{f_\alpha : V_\alpha \longrightarrow X\}$ such that $\{f_\alpha : \delta V_\alpha \longrightarrow X\}$ form a basis of the odd torsion subgroup of

$$\Omega_{4*-1}(X) \otimes_{\Omega_*} \mathbb{Z}_{(\text{odd})} \cong K_{-1}(X)\ .$$

(iv) Choose a collection $C$ of singular closed oriented $4i$-dimensional manifolds $\{g_s : M_s \longrightarrow X\}$ such that in

[1] For definition see remarks before Theorem 2′.

$$\begin{array}{ccc}
\Omega_{4*}(X) & \xrightarrow{S_* = \text{fund. class}} & \bigoplus_{i>0} H_{4i}(X;\mathbb{Z}_{(2)})/\text{Torsion} \\
\Big\downarrow {\scriptstyle I_* = \text{natural projection}} & & \Big\downarrow {\scriptstyle \text{inclusion}} \\
\Omega_{4*}(X) \otimes_{\Omega_*} \mathbb{Z}_{(\text{odd})}/\text{Torsion} & \xrightarrow{ph_*} & \bigoplus_{i>0} H_{4i}(X;\mathbb{Q})
\end{array}$$

$S_*C$ and $I_*C$ form bases.

I. If we assign "splitting obstructions" in $\mathbb{Z}_{p^r}$ or $\mathbb{Z}$ to each manifold in group (iii) or (iv) we define a collection of homomorphisms

$$\phi_n \ : \ K_0(X;\mathbb{Z}_n) \longrightarrow \mathbb{Z}_n \ , \ n \text{ odd or zero} .$$

The collection $\{\phi_n\}$ defines a unique element in $\sigma \in K^0(X)$, $\sigma : X \longrightarrow BO_{(\text{odd})}$.

The commutativity[1] of

$$\begin{array}{ccccc}
K_0(BO_{(\text{odd})};\mathbb{Z}_n) & & \xrightarrow[\text{eval. of id.}]{e_n(\text{id})} & & \mathbb{Z}_n \\
 & \nwarrow {\scriptstyle \sigma_*} & & \nearrow {\scriptstyle \phi_n} & \\
{\scriptstyle p_{(\text{odd})}} \Big\downarrow \cong & & K_0(X;\mathbb{Z}_n) & & \Big\uparrow {\scriptstyle \mathcal{S}\otimes} \\
K_0(F/PL;\mathbb{Z}_n) & & \xrightarrow{\cong} & & \Omega_{4*}(F/PL;\mathbb{Z}_n) \otimes_{\Omega_*} \mathbb{Z}_{(\text{odd})}
\end{array}$$

implies any lifting of $\sigma$

$$\begin{array}{ccc}
 & & F/PL \\
 & \overset{\overline{\sigma}}{\nearrow} & \Big\downarrow {\scriptstyle p_{(\text{odd})}} \\
X & \xrightarrow[\sigma]{} & BO_{(\text{odd})}
\end{array}$$

will have the desired splitting obstructions on these components.

II. If we give splitting obstructions for the 2 and 4 dimensional components, we can construct a homomorphism $H_2(M;\mathbb{Z}_2) \longrightarrow \mathbb{Z}_2$ using the given values on $A$ and the values (reduced mod 2) on the non-spin components of dimension 4 to obtain a cohomology class $u \in H^2(M;\mathbb{Z}_2)$ such that $\delta Sq^2 u = 0$.

[1] The outer commutativity is clear since for $Y = F/PL_{(\text{odd})}$ or $BO_{(\text{odd})}$ we have $K_0(Y;\mathbb{Z}_n) = K_0(Y) \otimes \mathbb{Z}_n$.

$u$ defines a map $X \longrightarrow K(\mathbb{Z}_2, 2)$ which may be lifted to $K(\mathbb{Z}_2, 2) \times_{\delta Sq^2} K(\mathbb{Z}_{(2)}, 4)$. We alter this lifting by a map $f : X \longrightarrow K(\mathbb{Z}_{(2)}, 4)$ to obtain the desired splitting invariants in dimension 2 and 4.

The splitting obstructions in the other dimensions may be obtained by mapping into the appropriate $K(\mathbb{Z}_{(2)}, 4i)$ or $K(\mathbb{Z}_2, 4i+2)$ independently. We obtain

$$\beta : X \longrightarrow \prod .$$

$\sigma$ and $\beta$ determine a unique map $f : X \longrightarrow F/PL$ with the desired splitting obstructions.

### Discussion of the characteristic variety

We can replace any component $g : N \longrightarrow X$ of the characteristic variety constructed for Theorem 4′ by $gp_2 : \mathbb{C} \times N \longrightarrow X$ if Index $\mathbb{C} = \pm 1$ and $N$ is not a $\mathbb{Z}$ or $\mathbb{Z}_{2^r}$ manifold of dimension 4. The new variety is still characteristic for $X$.

For determining whether two $F/PL$-bundles are the same we may further replace the four dimensional $\mathbb{Z}$-components $N$ by $\mathbb{C} \times N$. (The realization property is then disturbed however).

We cannot replace the 4-dimensional $\mathbb{Z}_{2^r}$-components by higher dimensional components because we thereby lose the delicate property that the splitting invariant is well defined modulo $2^{r+1}$ on these components.

Thus in either case the characteristic variety has two "parts" – one of dimension four and one of infinite (or stable) dimension.

The ability to stabilize is the real reason why only $\mathbb{Z}_n$-*manifolds* appear in Theorem 4′ and not varieties of the more complicated type discussed earlier (for the study of odd torsion). Such a variety $\times \mathbb{CP}^n$ is cobordant to a non-singular manifold.

The $\mathbb{Z}_n$-manifolds with singularities can be used to describe $F/PL$-bundles over $X$ together with filtrations (the highest skeleton over which the bundle is trivial).

## III. The Hauptvermutung

We can apply the *first* part of the *characteristic variety theorem* for $F/PL$ to study homeomorphisms.

**Theorem H.** *Let $h : (L, \partial L) \longrightarrow (M, \partial M)$ be a homeomorphism and $\theta_h : M \longrightarrow F/PL$ be the characteristic $F/PL$-bundle for $h$. Then there is only one possible non-zero obstruction to the triviality of $\theta_h$, an element of order 2 in $H^4(M; \mathbb{Z})$.*

Then if $H_3(M;\mathbb{Z})$ has no 2-torsion we have:

**Corollary 1.** *If $\pi_1(M) = \pi_1(\text{each component of } \partial M) = 0$ and $\dim M \geq 6$, then $h : (L, \partial L) \longrightarrow (M, \partial M)$ is homotopic to a PL-homeomorphism.*

**Corollary 2.** *In the non-simply-connected case, any dimension, we have that*

$$h \times \mathrm{id}_{\mathbb{R}^N} \ : \ (L, \partial L) \times \mathbb{R}^N \longrightarrow (M, \partial M) \times \mathbb{R}^N$$

*is properly homotopic to a PL-homeomorphism. We may take $N = 3$.*

**Corollary 3.** *The localizing projections for the natural map $H : TOP/PL \longrightarrow F/PL$ satisfy*

$$\begin{array}{ccc} (*) & \xrightarrow{\ Pt\ } & (F/PL)_{(\mathrm{odd})} \\ {\scriptstyle Pt}\uparrow & & \uparrow{\scriptstyle p_{\mathrm{odd}}} \\ TOP/PL & \xrightarrow{\ H\ } & F/PL \\ {\scriptstyle \Theta}\downarrow & & \downarrow{\scriptstyle p_{(2)}} \\ K(\mathbb{Z}_{(2)}, 4) & \xrightarrow[\text{inclusion}]{} & (F/PL)_{(2)} \end{array}$$

*where $\Theta$ is an h-map and has order 2.*
*($\mathbb{Z}_{(2)} = \mathbb{Z}[\frac{1}{3}, \frac{1}{5}, \ldots, \frac{1}{p_i}, \ldots]$, $p_i$ the $i^{\text{th}}$ odd prime).*[1]

**Corollary 4.** *Let $M$ be as in Corollary 1. The subgroup of $hT(M)$ generated by homeomorphisms $h : (L, \partial L) \longrightarrow (M, \partial M)$ is a $\mathbb{Z}_2$-module of dimension not exceeding the dimension of* $[\text{2-torsion } H_3(M;\mathbb{Z})] \otimes \mathbb{Z}_2$.

**Proof of Theorem H.**

Let $V$ be a characteristic variety for $M$. Replace each component $N$ of $V$ by $\mathbb{CP}^4 \times N$. If $M$ is simply connected, we may use the splitting theorem of Novikov [N1] to see that the splitting invariant of $h$ along $\mathbb{CP}^4 \times V$ equals zero.[2] This proves Theorem H in the simply connected case.

If $M$ is a general manifold we use a strengthened version of Novikov's splitting theorem (originally proved to treat the manifolds with singularities described

---

[1] We are considering the spaces of Corollary 3 as being defined by functors on the category of finite $CW$ complexes.

[2] $V$ may be chosen in this case so that $\pi_1(V) = \pi_1(\delta V) = 0$.

above).

**Definition.** A **manifold complex** is a polyhedron constructed inductively by attaching an $n$-dimensional $PL$-manifold to the previously constructed $(n-1)$-dimensional polyhedron along the boundary of the manifold which is embedded. The components of the $n$-manifold are the **$n$-cells** of the manifold complex.

**Splitting Lemma.** *Let $K$ be a manifold complex whose "cells" have dimension $\geq 5$ and free abelian fundamental group. Let $t : E \longrightarrow K \times \mathbb{R}^n$ be a topological trivialization of the piecewise linear $\mathbb{R}^n$-bundle $E$ over $K$. Then $t$ is properly homotopic to a map which is transverse regular to $K \times 0 \subset K \times \mathbb{R}^n$ and such that*

$$t| \; : \; t^{-1}(K \times 0) \longrightarrow K \times 0$$

*is a cell-wise homotopy equivalence of manifold complexes.*

**Proof of Splitting Lemma.** Assume first that $K$ has one cell. Consider

$$T^{n-1} \;=\; S^1 \times \ldots \times S^1 \; (n-1 \text{ factors}) \subset \mathbb{R}^n$$

and let

$$W \;=\; t^{-1}(K \times T^{n-1} \times \mathbb{R})\,.$$

Then $t_0 : W \longrightarrow K \times T^{n-1} \times \mathbb{R}$ is a *proper homotopy equivalence.*[1] We may apply Siebenmann's Thesis [S] to split $t_0$, namely we find a $PL$-homeomorphism $W \cong W_1 \times \mathbb{R}$ and a map $t_1 : W \longrightarrow K \times T^{n-1}$ so that

$$\begin{array}{ccc} W & \xrightarrow{\quad t_0 \quad} & K \times T^{n-1} \times \mathbb{R} \\ \cong \big\downarrow & \nearrow_{t_1 \times \mathrm{id}_{\mathbb{R}}} & \\ W_1 \times \mathbb{R} & & \end{array}$$

is properly homotopy commutative.

We then apply Farrell's fibring theorem [F] to deform

$$W_1 \xrightarrow{t_1} K \times T^{n-1} \xrightarrow{\text{last factor}} S^1$$

[1] This is the only place "homeomorphism" is used in the proof of the Hauptvermutung.

to a fibration and thus split $t_1$. Then we have a diagram

$$\begin{array}{ccc} W_2 & \xrightarrow{\quad t_2 \quad} & K \times T^{n-2} \\ \subset \Big\downarrow & & \Big\downarrow \subset \\ W_1 & \xrightarrow{\quad t_1' \quad} & K \times T^{n-1} \\ & \searrow \quad \swarrow & \\ & S^1 & \end{array}$$

where $t_1'$ is transverse regular to $K \times T^{n-2}$ and $t_2$ is a homotopy equivalence.

We similarly split $t_2$ and find a $t_3$, etc. Finally, after $n$ steps we obtain the desired splitting of $t$.

Now each of the above steps is relative (Siebenmann's $M \times \mathbb{R}$ theorem and Farrell's Fibring Theorem).

The desired splitting over a manifold complex may then be constructed inductively over the "cells". The only (and very crucial) requirement is that each manifold encountered has dimension $\geq 5$ and free abelian $\pi_1$.

**Proof of Theorem H (contd.)**

We may assume that $\theta_h : E \longrightarrow M \times \mathbb{R}^n$ is a topological bundle map (by increasing $n$ if necessary).

Now notice that (any $\mathbb{Z}_n$-manifold)$\times \mathbb{C}\mathbb{P}^4$ has the structure of a manifold complex satisfying the hypotheses of the *Splitting Lemma*. Thus the splitting invariant of $\theta_h$ along (characteristic variety)$\times \mathbb{C}\mathbb{P}^4$ is zero.

**Proof of Corollaries.**

*Corollary 1* follows from Theorem 3.

*Corollary 2* follows from the definition of $\theta_h$.

*Corollary 3* follows from Theorem 4.

*Corollary 4* follows from Theorems 5, 3 and 4.

Lashof and Rothenberg [LR] have proved the Hauptvermutung for 3–connected manifolds by deforming the 3–connective covering of

$$H : TOP/PL \longrightarrow F/PL$$

to zero. The argument is somewhat like that of the splitting lemma.

## Application to complex projective space

We will apply the general theory to the special case of complex projective space. We choose this example because (a) the results have immediate applications to the theory of free $S^1$-actions on homotopy spheres, (b) $\mathbb{CP}^n$ is interesting enough to illustrate certain complications in the theory, and finally (c) certain simplifications occur to make the theory especially effective in this case.

We illustrate the last point first. We assume $n > 2$ throughout.

**Theorem 8.** (i) *Any self-homotopy equivalence of $\mathbb{CP}^n$ is homotopic to the identity or the conjugation.*
(ii) *Any self-piecewise linear homeomorphism of $\mathbb{CP}^n$ is weakly isotopic to the identity or the conjugation.*

**Definition 4.** If $M$ is homotopy equivalent to $\mathbb{CP}^n$, we call a generator of $H^2(M;\mathbb{Z})$ **a *c*-orientation of** $M$.

**Corollary 1.** *The group of concordance classes of h-triangulations of $\mathbb{CP}^n$ is canonically isomorphic to the set of PL-homeomorphism classes of piecewise linear homotopy $\mathbb{CP}^n$'s.*

**Corollary 2.** *The set of concordance classes of h-smoothings of $\mathbb{CP}^n$ is canonically isomorphic to the set of c-oriented diffeomorphism classes of smooth homotopy $\mathbb{CP}^n$'s.*

**Corollary 3.** *The group of concordance classes of smoothings of $\mathbb{CP}^n$ is canonically isomorphic to the set of c-oriented diffeomorphism classes of smooth manifolds homeomorphic (or PL-homeomorphic) to $\mathbb{CP}^n$.*

**Proof of Theorem 8.**

(i) Theorem 8 (i) follows from the fact that $\mathbb{CP}^n$ is the $(2n+1)$-skeleton of $K(\mathbb{Z},2) = \mathbb{CP}^\infty$.

(ii) Any $PL$-homeomorphism $P : \mathbb{CP}^n \longrightarrow \mathbb{CP}^n$ is homotopic to the identity or the conjugation by (i). Choose (mod $\mathbb{CP}^n \times I$) such a homotopy $H$ and try to deform it to a weak isotopy (mod $\mathbb{CP}^n \times \partial I$)

$$H : \mathbb{CP}^n \times I \longrightarrow \mathbb{CP}^n \times I .$$

Such a deformation is obstructed (according to Theorem 1) by cohomology classes in

$$H^*(\mathbb{C}\mathbb{P}^n \times (I, \partial I); P_*) \; .$$

But these groups are all zero. This proves Theorem 8 (ii).

**Proof of Corollaries.**

Corollaries 2 and 3 follow immediately from Theorem 8 (i) and (ii) and the definitions.

Corollary 1 follows from the additional fact that

$$\begin{array}{ccc} \mathbb{C}\mathbb{P}^n & \xrightarrow{\;f\;} & \\ {\scriptstyle \text{conjugation}}\big\downarrow & & F/PL \\ \mathbb{C}\mathbb{P}^n & \xrightarrow{\;f\;} & \end{array}$$

is homotopy commutative for any $f$. The corollaries show that the three groups

$$(\mathbb{C}\mathbb{P}^n, PL/O) \quad , \quad (\mathbb{C}\mathbb{P}^n, F/O) \quad , \quad \text{and} \quad (\mathbb{C}\mathbb{P}^n, F/PL)$$

solve the correct problems.

**Theorem 9.** *The characteristic variety of* $\mathbb{C}\mathbb{P}^n$ *may be taken to be*

$$V \;=\; \mathbb{C}\mathbb{P}^2 \cup \mathbb{C}\mathbb{P}^3 \cup \ldots \cup \mathbb{C}\mathbb{P}^{n-1} \longrightarrow \mathbb{C}\mathbb{P}^n \; .$$

Thus any $PL$-manifold $M$ homotopy equivalent to $\mathbb{C}\mathbb{P}^n$ is determined uniquely by choosing any homotopy equivalence $g : M \longrightarrow \mathbb{C}\mathbb{P}^n$ and calculating the splitting invariant of $g$ along $V$. Furthermore *all such invariants are realizable.*

The set of $PL$-homeomorphism classes of such $M$ is therefore canonically isomorphic to

$$\begin{array}{lll} \mathbb{Z} & \text{for} & \mathbb{C}\mathbb{P}^3 \\ \mathbb{Z} \oplus \mathbb{Z}_2 & \text{for} & \mathbb{C}\mathbb{P}^4 \\ \mathbb{Z} \oplus \mathbb{Z}_2 \oplus \mathbb{Z} & \text{for} & \mathbb{C}\mathbb{P}^5 \\ \mathbb{Z} \oplus \mathbb{Z}_2 \oplus \mathbb{Z} \oplus \mathbb{Z}_2 & \text{for} & \mathbb{C}\mathbb{P}^6 \\ \vdots & \vdots & \vdots \end{array}$$

etc.

**Remark.** Any $\mathbb{C}\mathbb{P}^n$ admits a $c$-orientation reversing $PL$-homeomorphism, i.e. a piecewise linear conjugation. This follows from Corollary 1.

**Remark.** The characteristic variety $V$ does not contain $\mathbb{CP}^1$ because $A = \ker Sq^2 \subseteq H^2(\mathbb{CP}^n;\mathbb{Z}_2) = 0$. In fact, the splitting obstruction of $g$ along $\mathbb{CP}^1$ (i.e. Kervaire invariant of $g^{-1}(\mathbb{CP}^1)$) is just the splitting obstruction of $g$ along $\mathbb{CP}^2$ taken modulo 2 (i.e. $\frac{1}{8}(\text{Index }(g^{-1}\mathbb{CP}^2) - 1)(\text{modulo } 2)$).

**Remark.** $hT(\mathbb{CP}^n)$ has another group structure coming from the isomorphism

$$hT(\mathbb{CP}^n) \;\cong\; (\mathbb{CP}^n - \{\text{pt}\}, F/PL) \;\cong\; (\mathbb{CP}^{n-1}, F/PL)\ .$$

If we denote the $F/PL$-structure by $\otimes$ and the characteristic variety group structure by $+$ then the operation $a \circ b = (a \otimes b) - (a + b)$ is a multiplication.

The operations $\circ$ and $+$ make $hT(\mathbb{CP}^n)$ into a commutative associative ring.

The ring $hT(\mathbb{CP}^n) \otimes \mathbb{Z}_{(\text{odd})}$ has *one* generator $\eta$ obtained by *suspending* the additive generator of $hT(\mathbb{CP}^3) = \mathbb{Z}$. The elements $\eta, \eta^2, \ldots, \eta^{[\frac{n-1}{2}]}$ span $hT(\mathbb{CP}^n) \otimes \mathbb{Z}_{\text{odd}}$ additively.

**Remark.** A suspension map $\Sigma : hT(\mathbb{CP}^n) \longrightarrow hT(\mathbb{CP}^{n+1})$ is defined by

$$\begin{aligned}&(g : M \longrightarrow \mathbb{CP}^n) \;\longmapsto \\ &(\overline{g} : g^*\ (\text{line bundle}) \cup \text{cone on boundary} \longrightarrow \mathbb{CP}^{n+1})\ .\end{aligned}$$

If $H$ denotes the total space of the canonical $D^2$-bundle over $\mathbb{CP}^n$ then we have the diagram

$$\begin{array}{ccccc}
hT(\mathbb{CP}^n) & & \xrightarrow{\quad *\quad} & & hT(H) \\
 & \searrow^{\Sigma} & & \swarrow^{\cong} & \\
\theta \big\downarrow & & hT(\mathbb{CP}^{n+1}) & & \big\downarrow \theta_H\ \cong \\
hT(\mathbb{CP}^n) & & \xrightarrow[\cong]{\quad *\quad} & & (H, F/PL)
\end{array}$$

with $*$ given by the induced bundle.[1]

Thus the image of $\Sigma$ is isomorphic to

$$\text{image}(\theta) \;=\; \ker(\mathcal{S} : [\mathbb{CP}^n, F/PL] \longrightarrow P_n)$$

by Theorem 3.

**Corollary.** *An element in $hT(\mathbb{CP}^{n+1})$ is a suspension iff its top splitting invariant is zero.*

[1] $\theta_H$ is an isomorphism because $\pi_1(H) = \pi_1(\partial H) = 0$.

(Note: When we suspend elements of $hT(\mathbb{C}\mathbb{P}^n)$ we merely add zeroes to the string of splitting invariants.)

## Smoothing elements of $hT(\mathbb{C}\mathbb{P}^n)$

One interesting problem is to determine which elements of $hT(\mathbb{C}\mathbb{P}^n)$ are determined by smoothable manifolds.

For example,

$$(0,1) \in hT(\mathbb{C}\mathbb{P}^4)\ ,$$
$$(0,0,0,0,0,1) \in hT(\mathbb{C}\mathbb{P}^8)\ ,\ \text{and}$$
$$(0,0,0,0,0,0,0,0,0,0,0,0,0,0,1) \in hT(\mathbb{C}\mathbb{P}^{16})\ \text{are.}$$

In fact, these manifolds are stably $PL$-homeomorphic to the corresponding projective spaces.

A non-smoothable example is provided by

$$(0,0,0,0,1) \in hT(\mathbb{C}\mathbb{P}^6)\ .$$

*In fact any element of $hT(\mathbb{C}\mathbb{P}^n)$ with fifth invariant non-zero is non-smoothable.*[1]

Also any suspension of a non-smoothable homotopy $\mathbb{C}\mathbb{P}^n$ is likewise.

Understanding which $(4K+2)$-invariants are realizable by smooth manifolds is quite hard in general.

The corresponding problem for the $4K$-invariants is theoretically possible because of Adams' work on $J(\mathbb{C}\mathbb{P}^n)$.

This problem is further complicated by the fact that the set $hS(\mathbb{C}\mathbb{P}^n)$ has no natural group structure when $n$ is even. Theorem 3 asserts there are exact sequences

$$0 \longrightarrow hS(\mathbb{C}\mathbb{P}^{2n+1}) \longrightarrow (\mathbb{C}\mathbb{P}^{2n+1}, F/O) \xrightarrow{\mathcal{S}_1} \mathbb{Z}_2$$

$$0 \longrightarrow hS(\mathbb{C}\mathbb{P}^{2n}) \longrightarrow (\mathbb{C}\mathbb{P}^{2n}, F/O) \xrightarrow{\mathcal{S}_2} \mathbb{Z}\ .$$

$\mathcal{S}_1$ is a homomorphism, but $\mathcal{S}_2$ is not.

If we consider homotopy *almost smoothings* of $\mathbb{C}\mathbb{P}^n$ we do get a group

$$h^+S(\mathbb{C}\mathbb{P}^n)\ \cong\ (\mathbb{C}\mathbb{P}^{n-1}, F/O)\ ,$$

[1] This follows from the fact that the 10-dimensional Kervaire manifold is not a $PL$-boundary (mod 2).

and

$$h^+S(\mathbb{CP}^n) \xrightarrow{\text{natural}} hT(\mathbb{CP}^n)$$

$$hS(\mathbb{CP}^n) \xrightarrow{\text{natural}} hT(\mathbb{CP}^n)$$

are homomorphisms with the $\otimes$ structure on $hT(\mathbb{CP}^n)$.

It would bc interesting to describe all these group structures geometrically for $\mathbb{CP}^n$.

Corollary 3 asserts that the set of $c$-oriented equivalence classes of differentiable structures on $\mathbb{CP}^n$ is isomorphic to $[\mathbb{CP}^n, PL/O]$, a finite group.

We can calculate this group in another way if we ignore 2-torsion.

**Proposition**. $[\mathbb{CP}^n, PL/O]$ *is isomorphic to the zeroth stable cohomotopy group of* $\mathbb{CP}^n$ *modulo 2-torsion.*

**Proof**. We apply $(\mathbb{CP}^n, \ \ ) \otimes \mathbb{Z}[\frac{1}{2}]$ to the diagram

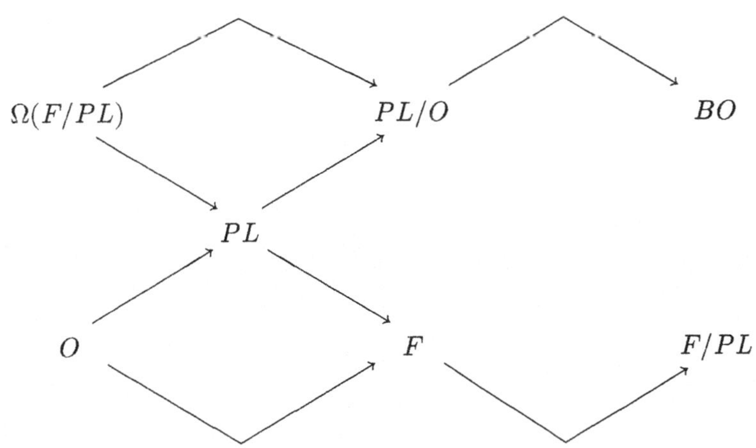


and obtain

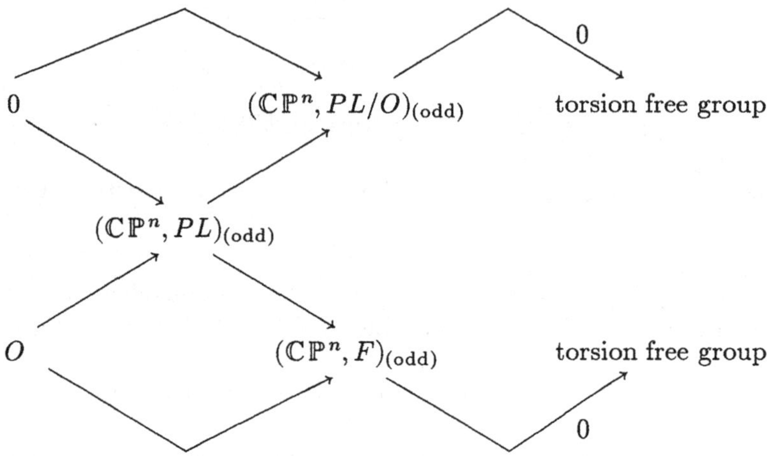


Thus

$$\begin{aligned}(\mathbb{CP}^n, PL/O)_{(\text{odd})} &\cong (\mathbb{CP}^n, PL)_{(\text{odd})} \\ &\cong (\mathbb{CP}^n, F)_{(\text{odd})} \equiv \pi_s^0(\mathbb{CP}^n)_{(\text{odd})} \, .\end{aligned}$$

## References


[A1] Anderson, D., **Duality in $K$-theory**, (to appear)

[AT] Araki, S. and Toda, S., **Multiplicative structures in mod $q$ Cohomology Theory I.**, Osaka J. Math. 2, 71–115, 1965.

[B1] Brown, E.H., **Cohomology Theories**, Annals of Math. 75, 467–484, 1962.

[CF1] Conner, P. and Floyd, E., **Differentiable Periodic Maps**, Springer Verlag, 1963.

[CF2] Conner, P. and Floyd, E., **Relation Between Cobordism and $K$-Theories**, Springer Verlag, 1967.

[F] Farrell, T., Thesis, Yale University, 1967

[KM] Kervaire, M. and Milnor, J., **Groups of Homotopy Spheres**, Annals of Math. 77, 504–537, 1963.

[LR] Lashof, R. and Rothenberg, M. (to appear)

[N1] Novikov, S.P., **On manifolds with free abelian fundamental group and applications (Pontrjagin classes, smoothings, high-dimensional knots)**, Izv. Akad. Nauk SSSR, ser. mat. 30, 208–246, 1966.

[N2] Novikov, S.P., **Written Report to International Congress**, Moscow, 1966.

[S] Siebenmann, L., Thesis, Princeton University, 1965.

[S1] Sullivan, Dennis, **Triangulating Homotopy Equivalences**, Thesis, Princeton University, 1966.

[W1] Wagoner, J., Thesis, Princeton University, 1966.


Princeton University
June 1967

## References

# The Princeton notes on the Hauptvermutung

by M.A.Armstrong, C.P.Rourke, G.E.Cooke

## Preface

The homotopy Hauptvermutung is the conjecture that a (topological) homeomorphism between two $PL$ (= piecewise linear) manifolds may be continuously deformed to a $PL$ homeomorphism.

These notes contain a proof, due to Casson and Sullivan, of the homotopy Hauptvermutung for simply connected manifolds under the hypothesis of 'no 2-torsion in $H^4$'. They were written in 1968 at the Institute for Advanced Study, Princeton and reissued in the Warwick Lecture Note Series in 1972. Nearly 25 years later there is still no other complete account available, hence their appearance in a more permanent form in this volume.

The connection with the subsequent solution of the isotopy Hauptvermutung by Kirby and Siebenmann [2, 3] is outlined in a coda. The two theories combine to give a fibration

$$K(\mathbb{Z}_2, 3) \simeq TOP/PL \longrightarrow G/PL \longrightarrow G/TOP \simeq \Omega^{4n}(G/PL)$$

and the following theorem.

**Theorem.** *Suppose that $h : Q \longrightarrow M$ is a (topological) homeomorphism between $PL$ manifolds of dimension at least five, whose restriction to $\partial M$ is $PL$. Then there is an obstruction $\theta \in H^3(M, \partial M; \mathbb{Z}_2)$ which vanishes if and only if $h$ is isotopic to a $PL$ homeomorphism keeping $\partial M$ fixed. If in addition $M$ is 1–connected then $h$ is homotopic to a $PL$ homeomorphism if and only if $\delta\theta \in H^4(M, \partial M; \mathbb{Z})$ is zero.*

When $M$ is not 1–connected the solution to the homotopy Hauptvermutung is bound to be more complicated (see the final remark in the coda).

More detail on the relationship of the results proved here with later results is to be found in the paper of Ranicki at the start of this volume.

The Princeton notes consist of three papers written by Armstrong, Rourke and Cooke, presented as three chapters, and a coda. The first chapter, written by Armstrong, gives an account of the Lashof-Rothenberg proof for 4–connected manifolds, and includes the 'canonical' Novikov splitting theorem used in the main argument. The second, by Rourke, contains the geometry of the main proof, and deals with simply connected manifolds which satisfy $H^3(M; \mathbb{Z}_2) = 0$. The treatment follows closely work of Casson on the global formulation of Sullivan theory.

This approach to the Hauptvermutung was the kernel of Casson's fellowship dissertation [1] and a sketch of this approach was communicated to Rourke by Sullivan in the Autumn of 1967. The remainder of the chapter contains an outline of an extension of the proof to the weakest hypothesis ($M$ simply connected and $H^4(M;\mathbb{Z})$ has no elements of order 2), and some side material on block bundles and relative Sullivan theory. The final chapter, written by Cooke, gives the details of the extension mentioned above. It contains part of Sullivan's analysis of the homotopy type of $G/PL$ and its application in this context. (The other part of this analysis is the verification that $G/PL$ and $BO$ have the same homotopy type 'at odd primes', see [5]). Sullivan's original arguments (outlined in [4, 5]) were based on his 'Characteristic Variety Theorem', and the present proof represents a considerable simplification on that approach.

Sadly George Cooke is no longer with us. We recall his friendship and this collaboration with much pleasure, and dedicate these notes to his memory.

M.A.A. (Durham), C.P.R. (Warwick)
January, 1996

**References**


[1] A. Casson, **Generalisations and applications of block bundles** Fellowship dissertation, Trinity College Library, Cambridge (1967). (Published in this volume).

[2] R. C. Kirby and L. C. Siebenmann, **On the triangulation of manifolds and the Hauptvermutung**, Bull. Amer. Math. Soc. 75 (1969), 742-749.

[3] R. C. Kirby and L. C. Siebenmann, **Foundational essays on topological manifolds, smoothings, and triangulations** Annals of Math. Studies 88, Princeton University Press (1977).

[4] D. P. Sullivan, **On the Hauptvermutung for manifolds**, Bull. Amer. Math. Soc. 73 (1967), 598-600.

[5] D. P. Sullivan, **Triangulating and smoothing homotopy equivalences and homeomorphisms. Geometric Topology Seminar Notes**, Princeton (1967). (Published in this volume).

# CHAPTER I

## The Hauptvermutung according to Lashof and Rothenberg

**By M. A. Armstrong**

### §1. Introduction

The aim of Chapter I is to prove the following result.

**(1.1) Theorem.** *Let $h : Q \longrightarrow M$ be a topological homeomorphism between two closed PL manifolds of dimension at least five. If $M$ is 4–connected, then $h$ is homotopic to a PL homeomorphism.*

The approach is due to Lashof and Rothenberg [11]. Our treatment differs only in that we write with the specialist less in mind, and prefer to emphasise the geometry throughout rather than enter a semi-simplicial setting. The theorem can be refined, but we shall examine only the version given above. Stronger results, due to Casson and Sullivan, are presented in Chapters II and III written by Rourke and Cooke.

In this introduction we shall present a bird's eye view of the proof of (1.1), referring the reader to later sections for more detail. Suppose then that $M$ and $Q$ are $PL$ manifolds, and that we are presented with a (topological) homeomorphism $h$ from $Q$ to $M$. We shall assume throughout that our manifolds are closed (compact without boundary) and 4–connected. The first step is to use $h$ to construct a $PL$ $\mathbb{R}^k$-bundle over $Q$, and a topological trivialization of this bundle. Second, by reference to Browder-Novikov theory, we show that if the given trivialization is properly homotopic to a $PL$ trivialization, then $h$ is homotopic to a $PL$ homeomorphism. The problem of homotoping the topological trivialization to a $PL$ trivialization will then occupy the remainder of the argument. Use of Browder-Novikov surgery necessitates the simple connectivity of our manifolds; the solution of the trivialization problem will require that the manifolds be 4–connected.

Identify $M$ with $M \times \{0\} \subseteq M \times \mathbb{R}^n$ for some large integer $n$, and think of $h$ as a (topological) embedding of $Q$ in $M \times \mathbb{R}^n$. If we are in the stable range, in other words if $n$ is at least $m+2$, a result of Gluck [6] provides an ambient isotopy $\{H_t\}$ of $M \times \mathbb{R}^n$ which moves $h$ to a $PL$ embedding

$$e = H_1 h : Q \longrightarrow M \times \mathbb{R}^n .$$

Further, in this range, work of Haefliger and Wall [7] shows that the new embedding has a $PL$ normal disc bundle. Taking the pullback gives a $PL$ $n$-disc bundle over

$Q$ and an extension of $e$ to a $PL$ embedding $e : E \longrightarrow M \times \mathbb{R}^n$ of its total space onto a regular neighbourhood $V$ of $e(Q)$ in $M \times \mathbb{R}^n$. Let $E \overset{\pi}{\longrightarrow} Q$ denote the associated $\mathbb{R}^n$-bundle. Choose a closed $n$-dimensional disc $D \subseteq \mathbb{R}^n$ centred on the origin, and of sufficiently large radius so that $V$ is contained in the interior of the 'tube' $M \times D$. Then $M \times D \backslash \mathrm{int}(V)$ is an $h$-cobordism between $M \times \partial D$ and $\partial V$ and, by the $h$-cobordism theorem, this region is $PL$ homeomorphic to the product $\partial V \times [0, 1]$. Therefore we may assume that $e : E \longrightarrow M \times \mathbb{R}^n$ is onto.

So far we have produced the following diagram

$$\begin{array}{ccc} E & \xrightarrow{\quad e \quad} & M \times \mathbb{R}^n \\ {\scriptstyle \pi}\big\downarrow & & \big\downarrow{\scriptstyle p_1} \\ Q & \xrightarrow{\quad h \quad} & M \end{array}$$

which commutes up to homotopy, with $e : E \longrightarrow M \times \mathbb{R}^n$ a $PL$ homeomorphism, and $\pi : E \longrightarrow Q$ a $PL$ $\mathbb{R}^n$-bundle. We claim that $e$ is stably isotopic to a topological bundle equivalence. Certainly the composition

$$H_1^{-1} e \ : \ E \longrightarrow M \times \mathbb{R}^n$$

provides a topological normal bundle for the embedding $h : Q \longrightarrow M \times \mathbb{R}^n$. On the other hand this embedding has a natural normal bundle given by

$$h \times 1 \ : \ Q \times \mathbb{R}^n \longrightarrow M \times \mathbb{R}^n \ .$$

These two are stably isotopic (see for example Hirsch [8] combined with Kister [9]). More precisely, if $r$ is at least $(m+1)^2 - 1$ then the associated normal bundles of

$$h \ : \ Q \longrightarrow M \times \mathbb{R}^n \times \mathbb{R}^r$$

are isotopic. Therefore

$$e \times 1 : E \times \mathbb{R}^r \longrightarrow M \times \mathbb{R}^n \times \mathbb{R}^r$$

is isotopic to a topological bundle equivalence. Let $g$ denote this equivalence, write

$$E' \ = \ E \times \mathbb{R}^r \ \ (k = n + r) \ ,$$

and consider

$$\begin{array}{ccccc} & \overset{g}{\frown} & & & \\ E' & \xrightarrow{\quad e \times 1 \quad} & M \times \mathbb{R}^k & \xleftarrow{\quad h \times 1 \quad} & Q \times \mathbb{R}^k \\ {\scriptstyle \pi p_1}\big\downarrow & & \big\downarrow{\scriptstyle p_1} & & \big\downarrow{\scriptstyle p_1} \\ Q & \xrightarrow{\quad h \quad} & M & \xleftarrow{\quad h \quad} & Q \end{array}$$

The composite

$$t = (h^{-1} \times 1)g : E' \longrightarrow Q \times \mathbb{R}^k$$

is topological trivialization of the $PL$ $\mathbb{R}^k$-bundle $\pi p_1 : E' \longrightarrow Q$.

Assume for the moment that $t$ is properly homotopic to a $PL$ trivialization. (We remind the reader that a proper map is one for which the inverse image of each compact set is always compact.) Then the inverse of this new trivialization followed by $e \times 1$ gives a $PL$ homeomorphism

$$f : Q \times \mathbb{R}^k \longrightarrow M \times \mathbb{R}^k$$

which is properly homotopic to

$$h \times 1 : Q \times \mathbb{R}^k \longrightarrow M \times \mathbb{R}^k .$$

Now let $\lambda : M \longrightarrow Q$ be a $PL$ map which is homotopic to $h^{-1}$, so that the composition

$$Q \times \mathbb{R}^k \xrightarrow{f} M \times \mathbb{R}^k \xrightarrow{\lambda \times 1} Q \times \mathbb{R}^k$$

is homotopic to the identity via a proper homotopy

$$F : Q \times \mathbb{R}^k \times I \longrightarrow \mathbb{R}^k .$$

Notice that both $F_0 = (\lambda \times 1)f$ and $F_1 = \mathrm{id} : Q \times \mathbb{R}^k \longrightarrow \mathbb{R}^k$ are $PL$ and transverse regular to the submanifold $Q \times \{0\}$ of $Q \times \mathbb{R}^k$. Also $F_0^{-1}(Q \times \{0\})$ is $PL$ homeomorphic to $M$, and $F_1^{-1}(Q \times \{0\})$ is the submanifold $Q \times \{0\} \times \{1\}$ of $Q \times \mathbb{R}^k \times \{1\}$. Using the relative simplicial approximation theorem of Zeeman [22] and the transverse regularity theorem of Williamson [20], we may assume without loss of generality that $F$ is itself $PL$ and transverse regular to $Q \times \{0\} \subseteq Q \times \mathbb{R}^k$. Let $W$ denote the compact manifold $F^{-1}(Q \times \{0\})$. Then $W$ is a proper submanifold of $Q \times \mathbb{R}^k \times I$ which has a trivial normal bundle (the pullback of the natural normal bundle of $Q \times \{0\} \subseteq Q \times \mathbb{R}^k$ under $F|_W$). Embed $Q \times \mathbb{R}^k$ in a sphere $S^N$ of high dimension, and extend this embedding in the obvious way to an embedding of $Q \times \mathbb{R}^k \times I$ in $S^{N+1}$. If $\nu_Q$ denotes the normal bundle of $Q \times \{0\}$ in $S^N$, and $\nu_W$ that of $W$ in $S^{N+1}$, then our map $F|_W : W \longrightarrow Q \times \{0\}$ extend to a bundle map $\nu_W \longrightarrow \nu_Q$.

Summarizing, we have produced a $PL$ manifold $W$, whose boundary consists of the disjoint union of $M$ and $Q$, and a $PL$ map $F : W \longrightarrow Q$ such that:

(i) $F|_Q$ is the identity;
(ii) $F|_M$ is a homotopy equivalence;
(iii) $F$ pulls back the stable $PL$ normal bundle of $Q$ to that of $W$.

In this situation we may apply the surgery results of Browder and Novikov [2, 3, 4, 14] to alter $W$ and $F$, though not $\partial W$ or $F|_{\partial W}$, until $F$ becomes a homotopy equivalence. The net result is an $h$-cobordism $W'$ between $M$ and $Q$, together with a deformation retraction $F' : W' \longrightarrow Q$. The $h$-cobordism theorem provides a $PL$

homeomorphism $G : M \times I \longrightarrow W'$, and $F'G$ is then a $PL$ homotopy between the maps

$$F_0G_0 \ , \ F_1G_1 \ : \ M \longrightarrow Q \ .$$

Now $F_0 : M \longrightarrow Q$ consists of a $PL$ automorphism of $M$, followed by a $PL$ map from $M$ to $Q$ that is homotopic to $h^{-1}$. Also, $F_1 : Q \longrightarrow Q$ is the identity map, and both of $G_0, G_1$ are $PL$ homeomorphisms. Therefore $h^{-1}$ is homotopic to a $PL$ homeomorphism. Consequently $h$ is also homotopic to a $PL$ homeomorphism. This completes our outline of the proof of Theorem 1.1.

**Remarks.** 1. If the dimension of $Q$ is even there is no obstruction to performing surgery. However, when the dimension is odd, there is an obstruction which must be killed and which, in the corresponding smooth situation, would only allow us to produce an $h$-cobordism between $M$ and the connected sum of $Q$ with an exotic sphere. Lack of exotic $PL$ spheres means that, in the $PL$ case, killing the surgery obstruction does not alter the boundary components of $W$.
2. In the terminology of Sullivan [18, 19] the $PL$ bundle $\pi : E \longrightarrow Q$ together with the fibre homotopy equivalence

$$\begin{array}{ccc} E & \xrightarrow{(h^{-1}\times 1)e} & Q \times \mathbb{R}^n \\ \pi \big\downarrow & & \big\downarrow p_1 \\ Q & = & Q \end{array}$$

is a characteristic $(F/PL)_n$-bundle for $h^{-1}$, and is classified by a homotopy class of maps from $Q$ to $F/PL$. Our work in stably moving $e$ to a topological bundle equivalence can be reinterpreted as factoring this class through $TOP/PL$. The final step (deforming the topological trivialization through fibre homotopy equivalences to a $PL$ trivialization) amounts to proving that the associated composite map

$$Q \longrightarrow TOP/PL \longrightarrow F/PL$$

is homotopically trivial. This will be the setting in the Chapters II, III by Rourke and Cooke.

Conversations with Colin Rourke were invaluable during the preparation of these notes, and I would like to thank him for his help.

## §2. Splitting theorems

At the end of §1 we were left with a $PL$ $\mathbb{R}^k$-bundle $E' \longrightarrow Q$, a topological trivialization $t : E' \longrightarrow Q \times \mathbb{R}^k$ and the problem of exhibiting a proper homotopy between $t$ and a $PL$ trivialization. Triangulate $Q$ in some way. We can now examine the restriction of the bundle to each simplex and try to push through an inductive argument. More precisely, let $\Delta$ be a simplex of the triangulation and $E'(\Delta)$ the part of the bundle over $\Delta$. Our problem reduces to that of constructing, inductively, a proper homotopy between $t|_{E'(\partial\Delta)}$ and a $PL$ bundle equivalence $E'(\partial\Delta)\longrightarrow\partial\Delta\times\mathbb{R}^k$, in such a way that it extends to one that moves $t|_{E'(\Delta)}$ to a $PL$ bundle equivalence $E'(\Delta)\longrightarrow\Delta\times\mathbb{R}^k$. This is the motivation for the 'splitting theorems' below.

Maps between bounded manifolds will, without further mention, be assumed to be maps of pairs (that is to say, they should carry boundary to boundary). Let $M$ be a compact topological manifold of dimension $m$, $W$ a $PL$ manifold of dimension $m+k$, and $h: W\longrightarrow M\times\mathbb{R}^k$ a proper homotopy equivalence.

**Definition.** A **splitting** for $h: W\longrightarrow M\times\mathbb{R}^k$ consists of a compact $PL$ manifold $N$, a $PL$ homeomorphism $s: N\times\mathbb{R}^k\longrightarrow W$ and a proper homotopy $\phi$ from $hs: N\times\mathbb{R}^k\longrightarrow M\times\mathbb{R}^k$ to $\lambda\times 1: N\times\mathbb{R}^k\longrightarrow M\times\mathbb{R}^k$, where $\lambda$ is the homotopy equivalence given by the composition

$$N \xrightarrow{\times 0} N\times\mathbb{R}^k \xrightarrow{hs} M\times\mathbb{R}^k \xrightarrow{\text{proj.}} M\ .$$

The splitting will be denoted by the ordered triple $(N,s,\phi)$. Remember, under our convention, $h$, $s$, and the proper homotopy $\phi$ all preserve boundaries. When $h$ has a splitting we shall simply say that $h$ **splits**. A splitting $(N',s',\phi')$ of $h|: \partial W\longrightarrow\partial M\times\mathbb{R}^k$ **extends** to one for $h$ if there is a splitting $(N,s,\phi)$ for $h: W\longrightarrow M\times\mathbb{R}^k$ such that $\partial N=N'$, $s|_{\partial N\times\mathbb{R}^k}=s'$ and $\phi|_{\partial N\times\mathbb{R}^k\times I}=\phi'$.

**(2.1) Splitting theorem.** *Let $W$ be a $PL$ manifold, $M$ a compact topological manifold and $h: W\longrightarrow M\times\mathbb{R}^k$ a homeomorphism. Then $h$ splits if $M$ is simply connected, and is either a closed manifold of dimension at least five, or has a simply connected boundary and dimension at least six.*

The proof of this theorem will occupy §4. We shall construct the splitting of $h$ using a very concrete construction due to Novikov, and we shall call a splitting a **Novikov splitting** if it arises in this way. There is a relative version of the theorem for Novikov splittings.

**(2.2) Relative splitting theorem.** *Let $W$ be a $PL$ manifold, $M$ a compact simply connected topological manifold of dimension at least five, and $h: W\longrightarrow M\times\mathbb{R}^k$ a homeomorphism. Then any Novikov splitting for $h|_{\partial W}$ extends to a Novikov splitting for $h$.*

A proof of this relative version is given in §5.

**Remarks.** (1) In our applications to the trivialization problem, the relative splitting theorem (2.2) will be applied in situations where $M$ is either a cell of dimension at least six, or $M$ is a cell of dimension five which is already supplied with a rather special splitting over the boundary. The Poincaré Conjecture will then tell us that the associated manifold $N$ is a $PL$ cell. Our hypothesis of 4–connectivity will enable us to avoid any reference to the splitting theorem over cells of dimension less than five.

(2) For $k = 1$ both theorems come directly from work of Siebenmann [16]. His arguments will not be repeated here, though his results are summarized in the next section. The manifold $N$ will occur in a very natural way as the boundary of a collar neighbourhood of an end of $M \times \mathbb{R}$. For higher values of $k$, ideas of Novikov allow us to produce a situation which is ripe for induction. Siebenmann's results are applied a second time in the inductive step.

## §3. Siebenmann's collaring theorems

In later sections we shall rely heavily on results from Siebenmann's thesis [16]. For completeness we sketch the necessary definitions and theorems. We remark that Siebenmann works entirely in the smooth category, however (as he notes) there are analogous $PL$ techniques, and we shall interpret all the results in $PL$ fashion.

An **end** $\mathcal{E}$ of a Hausdorff space $X$ is a collection of subsets which is maximal under the properties:

(i) Each member of $\mathcal{E}$ is a non-empty open connected set with compact frontier and non-compact closure;

(ii) If $A_1, A_2 \in \mathcal{E}$ then there exists $A_3 \in \mathcal{E}$ such that $A_3 \subseteq A_1 \cap A_2$.

(iii) The intersection of the closures of all the sets in $\mathcal{E}$ is empty.

A subset $U$ of $X$ is a **neighbourhood** of $\mathcal{E}$ if it contains some member of $\mathcal{E}$.

Our spaces are at worst locally finite simplicial complexes. For these one can show:

(i) The number of ends of $X$ is the least upper bound of the number of components of $X \backslash K$, where $K$ ranges over all finite subcomplexes of $X$.

(ii) The number of ends of $X$ is an invariant of the proper homotopy type of $X$.

A compact space has no ends; $\mathbb{R}$ has two ends and $\mathbb{R}^n$ has one end when $n \geq 2$; if $X$ is compact then $X \times \mathbb{R}$ has two ends; the universal covering space of the wedge of two circles has uncountably many ends. Think of a compact manifold with non-empty boundary. Removing a boundary component $M$ creates one end, and this end has neighbourhoods which are homeomorphic to $M \times [0, 1)$. Indeed the end has 'arbitrary small' neighbourhoods of this type, in the sense that every neighbourhood contains one of these so called collar neighbourhoods.

Let $W$ be a non-compact $PL$ manifold. A **collar** for an end $\mathcal{E}$ of $W$ is a connected $PL$ submanifold $V$ of $W$ which is a neighbourhood of $\mathcal{E}$, has compact boundary, and is $PL$ homeomorphic to $\partial V \times [0,1)$. In what follows we look for conditions on an end which guarantee the existence of a collar.

Given an end $\mathcal{E}$ of $W$, let $\{X_n\}$ be a sequence of path connected neighbourhoods of $\mathcal{E}$ whose closures have empty intersection. By selecting a base point $x_n$ from each $X_n$, and a path which joins $x_n$ to $x_{n+1}$ in $X_n$, we obtain an inverse system

$$\mathcal{S} \ : \ \pi_1(X_1, x_1) \xleftarrow{f_1} \pi_1(X_2, x_2) \xleftarrow{f_2} \cdots .$$

Following Siebenmann, we say that $\pi_1$ is **stable** at $\mathcal{E}$ if there is a sequence of neighbourhoods of this type for which the associated inverse system induces isomorphisms

$$\mathrm{im}(f_1) \xleftarrow{\cong} \mathrm{im}(f_2) \xleftarrow{\cong} \cdots .$$

When $\pi_1$ is stable at $\mathcal{E}$, define $\pi_1(\mathcal{E})$ to be the inverse limit of an inverse system $\mathcal{S}$ constructed as above. One must of course check that this definition is independent of all the choices involved.

Recall that a topological space $X$ is **dominated** by a finite complex $K$ if there are maps

$$K \underset{g}{\overset{f}{\rightleftarrows}} X$$

together with a homotopy

$$fg \simeq 1 \ : \ X \longrightarrow X .$$

Let $\mathcal{D}$ be the collection of all those spaces which are of the homotopy type of a $CW$ complex and dominated by a finite complex.

**Definition**. An end $\mathcal{E}$ of $W$ is **tame** if $\pi_1$ is stable at $\mathcal{E}$ and, in addition, there exist arbitrarily small neighbourhoods of $\mathcal{E}$ that lie in $\mathcal{D}$.

The **reduced projective class group** $\widetilde{K}_0(\mathbb{Z}[G])$ is the abelian group of stable isomorphism classes of finitely generated projective $\mathbb{Z}[G]$-modules.

**(3.1) The collaring theorem**. *Let $\mathcal{E}$ be a tame end of a $PL$ manifold which has compact boundary and dimension at least six. There is an obstruction in $\widetilde{K}_0(\mathbb{Z}[\pi_1(\mathcal{E})])$ which vanishes if and only if $\mathcal{E}$ has a collar.*

The corresponding relative version involves the ends of a $PL$ manifold $W$ whose boundary is $PL$ homeomorphic to the interior of a compact $PL$ manifold. (So in particular the ends of its boundary all have collars.) A **collar** for an end $\mathcal{E}$

of $W$ is now connected $PL$ submanifold neighbourhood $V$ of $\mathcal{E}$ such that:

(i) The frontier $bV$ of $V$ in $W$ is a compact $PL$ submanifold of $W$ (this frontier may itself have a boundary); and
(ii) $V$ is $PL$ homeomorphic to $bV \times [0, 1)$.

**(3.2) Relative collaring theorem.** *Let $\mathcal{E}$ be a tame end of a PL manifold which has dimension at least six, and whose boundary is PL homeomorphic to the interior of a compact PL manifold. Then $\mathcal{E}$ has a collar provided an obstruction in $\widetilde{K}_0(\mathbb{Z}[\pi_1(\mathcal{E})])$ vanishes. Further, the collar of $\mathcal{E}$ can be chosen to agree with any preassigned collars of those ends of $\partial W$ which are 'contained' in $\mathcal{E}$.*

**Remarks on the proof of (3.1).** A tame end is always isolated (in the sense that it has a neighbourhood which is not a neighbourhood of any other end), and its fundamental group is finitely presented. Given a tame end $\mathcal{E}$ of $W$, it is easy to produce a neighbourhood $V$ of $\mathcal{E}$ which is a connected $PL$ manifold having compact boundary and only one end. The idea is then to modify $V$ so that the inclusion of $\partial V$ in $V$ becomes a homotopy equivalence, when $V$ must be a collar by Stallings [17]. Preliminary modifications ensure that:

(i) $\partial V$ is connected,
(ii) the homomorphisms $\pi_1(\mathcal{E}) \longrightarrow \pi_1(V)$, and $\pi_1(\partial V) \longrightarrow \pi_1(V)$ induced by inclusion are isomorphisms and
(iii) the homology groups $H_i(\widetilde{V}, \widetilde{\partial V})$ are zero for $i \neq n-2$, where $n = \dim(W)$.

Here $\widetilde{V}$ denotes the universal covering space of $V$ and, by (ii), the part of $\widetilde{V}$ which sits over $\partial V$ is precisely the universal cover $\widetilde{\partial V}$ of $\partial V$. At this stage $H_{n-2}(\widetilde{V}, \widetilde{\partial V})$ turns out to be a finitely generated projective $\mathbb{Z}[\pi_1(\mathcal{E})]$-module. The class of this module in $\widetilde{K}_0(\mathbb{Z}[\pi_1(\mathcal{E})])$ is the obstruction mentioned in the statement of (3.1). When this module is stably free we can modify $V$ further so that $H_*(\widetilde{V}, \widetilde{\partial V})$ is zero, and the inclusion of $\partial V$ in $V$ is then a homotopy equivalence.

The following result will be needed later. Let $M$ be a compact topological manifold of dimension $m$, $W$ a $PL$ manifold of dimension $m+1$, and $h : W \longrightarrow M \times \mathbb{R}$ a proper homotopy equivalence of pairs.

**(3.3) Theorem.** *The ends of $W$ are tame.*

**Proof.** Since $h$ is a proper homotopy equivalence $W$ has exactly two ends. Let $g$ be a proper homotopy inverse for $h$, and $\mathcal{E}$ the end whose neighbourhoods contain sets of the form $g(M \times [t, \infty))$.

(a) Given a path connected neighbourhood $X$ of $\mathcal{E}$, choose $t$ so that $g(M \times [t, \infty))$ is contained in $X$. Write $\alpha(X)$ for the homomorphism from $\pi_1(M)$ to $\pi_1(X)$ induced by the composite map

$$M \xrightarrow{\times t} M \times [t, \infty) \xrightarrow{g} X ,$$

and note that $\alpha(X)$ is a monomorphism because $hg$ is homotopic to the identity map of $M \times \mathbb{R}$.

Begin with a path connected neighbourhood $X_1$ of $\mathcal{E}$. If $F$ is a proper homotopy from $gh$ to the identity map of $W$, choose a path connected neighbourhood $X_2$ of $E$ which lies in the interior of $X_1$ and satisfies $F(X_2 \times I) \subseteq X_1$. If $f_1 : \pi_1(X_2) \longrightarrow \pi_1(X_1)$ is induced by inclusion we have a commutative diagram

$$\begin{array}{ccc} \pi_1(X_2) & \xrightarrow{\quad f_1 \quad} & \pi_1(X_1) \\ {\scriptstyle \alpha(X_2)} \nwarrow & & \nearrow {\scriptstyle \alpha(X_1)} \\ & \pi_1(M) & \end{array}$$

We claim that $\text{im}(\alpha(X_1)) = \text{im}(f_1)$, so that $\alpha(X_1)$ is an isomorphism from $\pi_1(M)$ to $\text{im}(f_1) \subseteq \text{im}(\alpha(X_1))$. We now select $X_3$ in the interior of $X_2$ with the property $F(X_3 \times I) \subseteq X_2$, and so on. The inverse system

$$\mathcal{S} : X_1 \xleftarrow{f_1} X_2 \xleftarrow{f_2} X_3 \longleftarrow \cdots$$

then shows that $\pi_1$ is stable at $\mathcal{E}$, and that $\pi_1(\mathcal{E})$ is isomorphic to $\pi_1(M)$.

(b) We quote the following lemma from Siebenmann [16].

**Lemma.** *Let $Z$ be connected CW complex which is the union of two connected sub-complexes $Z_1, Z_2$. If $Z_1 \cap Z_2, Z \in \mathcal{D}$, and if both $\pi_1(Z_1), \pi_1(Z_2)$ are retracts of $\pi_1(Z)$, then $Z_1, Z_2 \in \mathcal{D}$.*

From part (a) we know that $\pi_1$ is stable at $\mathcal{E}$, and that $\pi_1(\mathcal{E})$ is finitely presented. To see the latter, remember that $\pi_1(\mathcal{E}) \cong \pi_1(M)$ and that $M$ is a compact topological manifold, and therefore dominated by a finite complex. Hence $\pi_1(M)$ is a retract of a finitely presented group and is itself finitely presented. Assume for simplicity that $M$ is closed. Given a neighbourhood $X$ of $\mathcal{E}$, Siebenmann's methods allow us to construct a connected $PL$ submanifold neighbourhood $V$ inside $X$ such that the homomorphisms $\pi_1(\mathcal{E}) \longrightarrow \pi_1(V)$, $\pi_1(\partial V) \longrightarrow \pi_1(V)$ induced by inclusion are both isomorphisms. Then $\pi_1(V)$ and $\pi_1(W \backslash \text{int}(V))$ are both isomorphic to $\pi_1(W)$. To complete the proof of (3.3) we simply apply the lemma, taking $Z = W$, $Z_1 = V$ and $Z_2 = W \backslash \text{int}(W)$.

One can define $\pi_r$ to be stable at $\mathcal{E}$ in exactly the same way as for $\pi_1$. Having done this the first part of the above proof is easily modified to give:

**(3.4) Addendum.** *If $\mathcal{E}$ is an end of $W$ then, for each $r$, $\pi_r$ is stable at $\mathcal{E}$ and*

$$\pi_r(\mathcal{E}) \cong \pi_r(M) \cong \pi_r(W) .$$

## §4. Proof of the splitting theorem

We consider the splitting theorem (2.1) in its simplest form. As before let $W$ be a $PL$ manifold, $M$ a closed simply connqcted topological manifold of dimension at least five and $h : W \longrightarrow M \times \mathbb{R}^k$ a homeomorphism. We must show that $h$ splits.

Let $T^k$ denote the $k$-dimensional torus (the cartesian product of $k$ copies of the circle), and let

$$D = \{(x_1, x_2, \ldots, x_k) \in \mathbb{R}^k \,\|\, |x_j| \leq 1, 1 \leq j \leq k\} .$$

Starting from an embedding of $S^1 \times \mathbb{R}$ in $\mathbb{R}^2$ we can inductively define embeddings

$$T^{k-1} \times \mathbb{R} \subseteq \mathbb{R}^k$$

for which the universal covering projection

$$e = \exp \times 1 : \mathbb{R}^{k-1} \times \mathbb{R} \longrightarrow T^{k-1} \times \mathbb{R}$$

is the identity on a neighbourhood of $D$. We leave the details to the reader.

If $P$ denotes $h^{-1}(M \times T^{k-1} \times \mathbb{R})$, then $P$ is an open subset of $W$ and therefore inherits a $PL$ structure from $W$. Write $h_1$ for the restriction of $h$ to $P$, and consider the pullback from :

$$\begin{array}{ccc} \widetilde{P} & \overset{\widetilde{h}_1}{\dashrightarrow} & M \times \mathbb{R}^k \\ {\scriptstyle p}\downarrow & & \downarrow{\scriptstyle 1 \times e} \\ P & \xrightarrow{h_1} & M \times T^{k-1} \times \mathbb{R} \end{array}$$

Since $M$ is simply connected, $\widetilde{P}$ is just the universal cover of $P$, and $p$ the associated covering projection. Let $i$ denote the inclusion map of $P$ in $W$.

**(4.1) Theorem.** *There is a $PL$ homeomorphism $d : \widetilde{P} \longrightarrow W$ such that:*
(i) *$d = ip$ on a neighbourhood of $\widetilde{h}_1^{-1}(M \times D)$, and*
(ii) *$hd$ is isotopic to $\widetilde{h}_1$ keeping a neighbourhood of $\widetilde{h}_1^{-1}(M \times D)$ fixed.*

**Remark.** In view of (4.1) we shall be able to restrict ourselves to the problem of splitting $\widetilde{h}_1 : \widetilde{P} \longrightarrow M \times \mathbb{R}^k$.

**Proof of 4.1**. Diagramatically we have

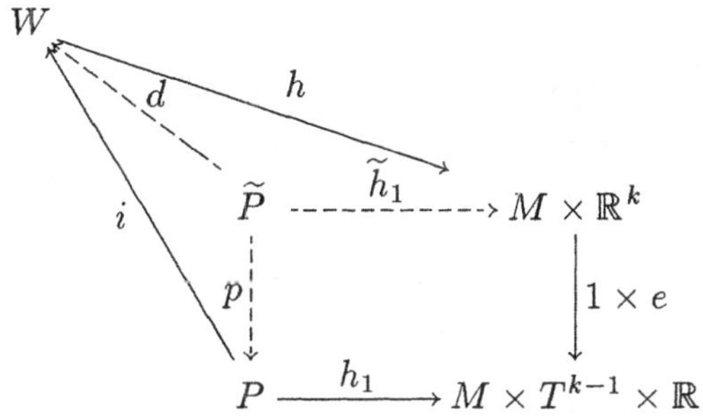


The map $1 \times e : M \times \mathbb{R}^k \longrightarrow M \times T^{k-1} \times \mathbb{R}$ is the identity on $M \times D_\epsilon$, for some $\epsilon > 1$, where

$$D_\epsilon = \{(x_1, x_2, \ldots, x_k) \in \mathbb{R}^k \,\|\, |x_j| \leq \epsilon, 1 \leq j \leq k\} .$$

Therefore $ip$ $PL$ embeds $\widetilde{h}_1^{-1}(M \times D_\epsilon)$ in $W$. Now $\widetilde{h}_1^{-1}(M \times D_\epsilon)$ has only one end; it is clearly tame and its fundamental group is trivial because $M$ is simply connected. The collaring theorem provides a compact $PL$ submanifold $B$ of $\widetilde{P}$ such that

$$\widetilde{h}_1^{-1}(M \times \mathrm{int}(D_\epsilon)) = B \cup \partial B \times [0,1) \quad , \quad \widetilde{h}_1^{-1}(M \times D) \subset \mathrm{int}(B) .$$

Consider the $PL$ manifold $\widetilde{P} \backslash \mathrm{int}(B)$. Again we have one simply connected end and, if $V$ is a collar of this end, the region $\widetilde{P} \backslash (\mathrm{int}(B) \cup \mathrm{int}(V))$ is an $h$-cobordism. Hence by Stallings [17] there is a $PL$ homeomorphism $\gamma : \widetilde{P} \longrightarrow B \cup \partial B \times [1,0)$ which is the identity on $B$. At this stage $ip\gamma : \widetilde{P} \longrightarrow W$ is a $PL$ embedding that agrees with $ip$ on $B$. By the same trick, applied this time in $W$, we can 'expand' $ip\gamma$ to provide a $PL$ homeomorphism $d : \widetilde{P} \longrightarrow W$ which satisfies (i).

To deal with property (ii) it is sufficient to show that

$$\psi = hd\widetilde{h}_1^{-1} : M \times \mathbb{R}^k \longrightarrow M \times \mathbb{R}^k$$

is isotopic to the identity keeping a neighbourhood of $M \times D$ fixed. Write

$$\psi(m,x) = (\psi^1(m,x), \psi^2(m,x))$$

and use the 'Alexander isotopy' defined by

$$\psi_0 = \text{identity} ,$$

$$\psi_t(m,x) = (\psi^1(m,tx), \frac{1}{t}\psi^2(m,tx)) \quad (0 < t \leq 1) .$$

This completes the proof of (4.1).

We make a couple of assertions concerning proper maps, leaving the reader to fill in the details.

**(4.2) Assertion**. *Let $A$ and $B$ be compact spaces. A map $f : A \times \mathbb{R}^n \longrightarrow B \times \mathbb{R}^n$*

*is proper if and only if given an arbitrarily large positive real number* $\epsilon$ *there is a positive* $\delta$ *such that* $|p_2 f(a,x)| > \epsilon$, *for all* $a \in A$ *and* $x \in \mathbb{R}^n$ *with* $|x| > \delta$.

**(4.3) Assertion.** *A bundle map between two bundles which have locally compact base spaces and a locally compact fibre is proper if and only if the corresponding map of base spaces is proper.*

**(4.4) Theorem.** *Let* $P$ *be a PL manifold of dimension* $m+r+1$ *and* $h : P \longrightarrow M \times T^r \times \mathbb{R}$ *a proper homotopy equivalence. Then* $h$ *splits.*

**Proof.** Since $h$ is a proper homotopy equivalence, $P$ has exactly two ends. Both are tame by (3.3), and their fundamental groups are free abelian of rank $r$. There is no obstruction to collaring because $\widetilde{K_0}(\mathbb{Z}[\mathbb{Z}^r])$ is the trivial group (see [1]). Let $U$ and $V$ be disjoint collars of the two ends. Addendum (3.4) can be used to see that $P\backslash(\text{int}(U) \cup \text{int}(V))$ is an $h$-cobordism, and therefore $P\backslash(\text{int}(U) \cup V)$ is $PL$ homeomorphic to $\partial U \times [0,1)$. But $P\backslash(\text{int}(U) \cup V)$ is also $PL$ homeomorphic to $P\backslash\text{int}(U)$. Collecting together this information we find there is a $PL$ homeomorphism $s : \partial U \times \mathbb{R} \longrightarrow P$. Let

$$g = hs : \partial U \times \mathbb{R} \longrightarrow M \times T^r \times \mathbb{R},$$

and write

$$g(u,x) = (g_1(u,x), g_2(u,x))$$

where $g_2(u,x) \in \mathbb{R}$. As $g$ is a proper map, the limit of $g_2(u,x)$ as $x$ tends to $+\infty$ is either $+\infty$ or $-\infty$ simultaneously for all $u \in U$. We assume $s$ chosen so as to give the positive limit. The map

$$\begin{aligned} \phi : \partial U \times \mathbb{R} \times [0,1] &\longrightarrow M \times T^r \times \mathbb{R}; \\ (u,x,t) &\longrightarrow (g_1(u,tx), (1-t)x + tg_2(u,x)) \end{aligned}$$

is a homotopy between $g$ and $\lambda \times \text{id}_{\mathbb{R}}$, where $\lambda(u) = g_1(u,0)$. Using (4.2) one easily checks that $\phi$ is a proper map. Therefore $(\partial U, s, \phi)$ splits $h$.

Returning to the terminology of 4.1 we obtain the next step in the proof of the splitting theorem (2.1):

**(4.5) Theorem.** *The homeomorphism* $\widetilde{h}_1 : \widetilde{P} \longrightarrow M \times \mathbb{R}^k$ *splits.*

**Proof.** Apply (4.4) repeatedly to construct a **tower** for $h_1 : P \longrightarrow M \times T^{k-1} \times \mathbb{R}$ as illustrated below.

$$\begin{array}{ccc}\overline{P}_k \times \mathbb{R} & \xrightarrow{h_k} & M \times \mathbb{R} \\ & & \downarrow \\ & & M \times S^1\end{array}$$

$$\cdots \quad \cdots \quad \cdots \quad \cdots \quad \cdots$$

$$\begin{array}{l}\overline{P}_2 \times \mathbb{R} \xrightarrow{s_2} P_2 \xrightarrow{h_2} M \times T^{k-2} \times \mathbb{R} \to M \times T^{k-2} \\ \qquad\qquad\qquad\qquad\qquad\qquad \downarrow \\ \overline{P}_1 \times \mathbb{R} \xrightarrow{s_1} P \xrightarrow{h_1} M \times T^{k-1} \times \mathbb{R} \longrightarrow M \times T^{k-1}\end{array}$$

We say a few words in case the reader starts operating at the wrong end of the diagram. Start with the homeomorphism $h_1 : P \longrightarrow M \times T^{k-1} \times \mathbb{R}$. Split this using (4.4) to obtain a compact $PL$ manifold $\overline{P}_1$, a $PL$ homeomorphism $s_1 : \overline{P}_1 \times \mathbb{R} \longrightarrow P$ and a proper homotopy $\phi^1$ from $h_1 s_1$ to $\lambda_1 \times 1$, where $\lambda_1$ is the homotopy equivalence given by the composite

$$\overline{P}_1 \xrightarrow{\times 0} \overline{P}_1 \times \mathbb{R} \xrightarrow{h_1 s_1} M \times T^{k-1} \times \mathbb{R} \xrightarrow{\text{proj.}} M \times T^{k-1} .$$

Now induce $h_2 : P_2 \longrightarrow M \times T^{k-2} \times \mathbb{R}$ as the pullback

$$\begin{array}{ccc} P_2 & \overset{h_2}{\dashrightarrow} & M \times T^{k-2} \times \mathbb{R} \\ \Big\downarrow & & \Big\downarrow \mathrm{id}_{M \times T^{k-2}} \times \exp \\ \overline{P}_1 & \xrightarrow{\lambda_1} & M \times T^{k-1} \end{array}$$

Then $h_2$ is a proper homotopy equivalence by Assertion (4.3). Split again using (4.4) to produce $(\overline{P}_2, s_2, \phi^2)$, and so on. The process terminates after $k$ steps.

For each $r$ let $\widetilde{P}_r$ denote the universal covering space of $P_r$. There are induced bundle maps

$$\begin{array}{ccccc}
\widetilde{P}_{r+1} & \overset{\widetilde{s}_r}{\dashrightarrow} & \widetilde{P}_r & \overset{\widetilde{h}_r}{\dashrightarrow} & M \times \mathbb{R}^{k-r+1} \\
\downarrow & & \downarrow & & \downarrow {\scriptstyle 1 \times \exp \times 1} \\
\overline{P}_r \times \mathbb{R} & \xrightarrow{s_r} & P_r & \xrightarrow{h_r} & M \times T^{k-r} \times \mathbb{R}
\end{array}$$

Note that

$$\widetilde{P}_{k+1} = \overline{P}_k \ , \ \widetilde{P}_k = P_k \ , \ \widetilde{s}_k = s_k \ , \ \widetilde{h}_k = h_k \ .$$

Let

$$N = \overline{P}_k = \widetilde{P}_{k+1} \ ,$$

and let $\widetilde{s}$ denote the composition

$$N \times \mathbb{R}^k \xrightarrow{\widetilde{s}_k \times 1} \widetilde{P}_k \times \mathbb{R}^{k-1} \xrightarrow{\widetilde{s}_{k-1} \times 1} \cdots \xrightarrow{\widetilde{s}_2 \times 1} \widetilde{P}_2 \times \mathbb{R} \xrightarrow{\widetilde{s}_1} \widetilde{P}$$

where $\widetilde{s}_r \times 1$ stands for $\widetilde{s}_r \times \mathrm{id}_{\mathbb{R}^{r-1}}$. Then $N$ is a compact $PL$ manifold and

$$\widetilde{s} : N \times \mathbb{R}^k \longrightarrow \widetilde{P}$$

is a $PL$ homeomorphism. We are left to construct a proper homotopy $\widetilde{\phi}$ between $\widetilde{h}_1\widetilde{s}$ and the usual product $\widetilde{\lambda} \times \mathrm{id}_{\mathbb{R}^k}$. For each $r$ the tower construction provides a proper homotopy $\phi^r$ from $h_r s_r$ to $\lambda_r \times \mathrm{id}_{\mathbb{R}}$. These lift to proper homotopies (use (4.3) again) from $\widetilde{h}_r\widetilde{s}_r$ to $\widetilde{h}_{r+1} \times \mathrm{id}_{\mathbb{R}}$, which in turn induce proper homotopies

$$\begin{aligned}
\widetilde{h}_1\widetilde{s} = \widetilde{h}_1\widetilde{s}_1(\widetilde{s}_2 \times 1)\cdots(\widetilde{s}_k \times 1) &\simeq (\widetilde{h}_2\widetilde{s}_2 \times 1)\cdots(\widetilde{s}_k \times 1) \\
&\simeq \cdots \\
&\simeq \widetilde{h}_k\widetilde{s}_k \times 1 \\
&\simeq \widetilde{\lambda} \times 1 \ .
\end{aligned}$$

If $\widetilde{\phi}$ denotes the composite proper homotopy from $\widetilde{h}_1\widetilde{s}$ to $\widetilde{\lambda} \times 1$, then $(N, \widetilde{s}, \widetilde{\phi})$ splits $\widetilde{h}_1$. This completes the proof of (4.4).

**Proof of the splitting theorem (2.1).** By (4.1) and (4.4) we have the following situation

$$N \times \mathbb{R}^k \xrightarrow{\widetilde{s}} \widetilde{P} \xrightarrow{d} W \xrightarrow{h} M \times \mathbb{R}^k$$

where $hd$ is isotopic to $\widetilde{h}_1 : \widetilde{P} \longrightarrow M \times \mathbb{R}^k$. Let

$$s = d\widetilde{s} : N \times \mathbb{R}^k \longrightarrow W ,$$

and construct a proper homotopy $\phi$ from $hs$ to $\lambda \times \mathrm{id}_{\mathbb{R}^k}$ as the composition

$$hs = hd\widetilde{s} \simeq \widetilde{h}_1\widetilde{s} \simeq \widetilde{\lambda} \times \mathrm{id}_{\mathbb{R}^k} \simeq \lambda \times \mathrm{id}_{\mathbb{R}^k} .$$

The triple $(N, s, \phi)$ is a splitting for $h : W \longrightarrow M \times \mathbb{R}^k$, as required.

We have proved the splitting theorem when $M$ is a closed simply connected manifold of dimension at least five. Exactly the same process goes through for compact, simply connected, manifolds of dimension at least six which have a simply connected boundary. All maps and homotopies must now preserve boundaries, and the relative collaring theorem is needed for the bounded analogues of (4.1) and (4.4).

## §5. Proof of the relative splitting theorem

Let $W$ be a $PL$ manifold, $M$ a compact topological manifold (which may have boundary), and $h : W \longrightarrow M \times \mathbb{R}^k$ a homeomorphism.

**Definition.** A splitting of $h$ is a **Novikov splitting** if it can be obtained by the construction presented in §4.

More precisely, a splitting $(N, s, \phi)$ of $h$ is a Novikov splitting if (keeping the previous notation) we can find a $PL$ homeomorphism $d : \widetilde{P} \longrightarrow W$ satisfying the hypotheses of (4.1), plus a tower for $h_1 : P \longrightarrow M \times T^{k-1} \times \mathbb{R}$, such that $N = \overline{P}_k$, $s = d\widetilde{s}$ and $\phi$ can be constructed from the tower homotopies and the isotopy of (4.1) in the manner described earlier.

Note that it makes sense to speak of a Novikov splitting for $h : W \longrightarrow M \times \mathbb{R}^k$ even when $M$ is not simply connected. Of course, in this case, the covering spaces involved are no longer universal coverings. For example, $\widetilde{P}$ becomes the cover of $P$ which corresponds to the subgroup $\pi_1(M) \lhd \pi_1(M \times T^{k-1} \times \mathbb{R})$.

In the special case where $M$ is a $PL$ manifold, and $h$ is a $PL$ homeomorphism, then $(M, h^{-1}, h^{-1} \times 1)$ is a splitting for $h$ and will be called the **natural splitting**.

It is a Novikov splitting. Just take $d = h^{-1}\widetilde{h}_1 : \widetilde{P} \longrightarrow W$ and use

$$\begin{aligned} \overline{P}_1 &= M \times T^{k-1} \ , \quad s_1 = h_1^{-1} \ , \\ P_r &= M \times T^{k-r} \times \mathbb{R} \ , \quad \overline{P}_r = M \times T^{k-r} \ , \\ h_r &= s_r = \text{identity} \ (r > 1) \ , \end{aligned}$$

as a tower for $h_1 : P \longrightarrow M \times T^{k-1} \times \mathbb{R}$.

For the remainder of this section we shall assume that $M$ is simply connected and has dimension at least five. Given a Novikov splitting for $h|_{\partial W}$, we must show that it extends to a Novikov splitting for $h$. There are two essential ingredients in the construction of a Novikov splitting, namely a suitable $PL$ homeomorphism $d : \widetilde{P} \longrightarrow W$ and a tower for $h_1 : P \longrightarrow M \times T^{k-1} \times \mathbb{R}$. We therefore need relative versions of (4.1) and (4.4).

**(5.1) Theorem**. *Suppose that $d' : \partial\widetilde{P} \longrightarrow \partial W$ is a $PL$ homeomorphism which satisfies:*

(i) *$d' = ip|_{\partial\widetilde{P}}$ on a neighbourhood of $\widetilde{h}_1^{-1}(\partial M \times D)$, and*

(ii) *$(h|_{\partial W})d'$ is isotopic to $\widetilde{h}_1|_{\partial\widetilde{P}}$ keeping a neighbourhood of $\widetilde{h}_1^{-1}(\partial M \times D)$ fixed.*

*Then there is a $PL$ homeomorphism $d : \widetilde{P} \longrightarrow W$ which satisfies* (i) *and* (ii) *of (4.1), such that $d|_{\partial\widetilde{P}} = d'$ and the isotopy of $hd$ extends that of $(h|_{\partial W})d'$.*

**Proof**. Proceeding essentially as in (4.1) we use the relative collaring theorem (3.2) to construct a $PL$ homeomorphism $\overline{d} : \widetilde{P} \longrightarrow W$ such that $d = ip$ on a neighbourhood of $h_1^{-1}(M \times D)$, and $hd$ is isotopic to $\widetilde{h}_1$ keeping a neighbourhood of $\widetilde{h}_1^{-1}(M \times D)$ fixed. Along the way we write

$$\widetilde{P} = B \cup bB \times [0,1) \ , \quad \partial\widetilde{P} = B' \cup \partial B' \times [0,1)$$

where $B$ is a compact $PL$ submanifold of $\widetilde{P}$ which meets $\partial\widetilde{P}$ transversally, $B' = B \cap \partial\widetilde{P}$, $bB$ is the frontier of $B$ in $\widetilde{P}$, and:

$$\begin{aligned} &\widetilde{h}_1^{-1}(M \times D) \subseteq \text{int}(B) \ , \\ &\overline{d} = ip \quad \text{on a neighbourhood of } B \ , \\ &\overline{d}|_{\partial\widetilde{P}} = d' \quad \text{on a neighbourhood of } B' \ . \end{aligned}$$

Since $\overline{d}|_{\partial B' \times [0,1)}$ , $d'|_{\partial B' \times [0,1)}$ are collars of $d'(\partial B')$ in $\partial W \backslash d'(\text{int}(B'))$, there is a $PL$ ambient isotopy of $\partial W$ which moves $\overline{d}|_{\partial\widetilde{P}}$ so as to agree with $d'$ whilst keeping $d'(B')$ fixed. Extend this ambient isotopy to an ambient isotopy $H$ of all of $W$ which keeps $\overline{d}(B)$ fixed, and let

$$d = H_1\overline{d} : \widetilde{P} \longrightarrow W \ .$$

Then by construction we have $d|_{\partial\widetilde{P}} = d'$ and $d = ip$ on a neighbourhood of $\widetilde{h}_1^{-1}(M \times D)$.

Take the given isotopy from $(h|_{\partial W})d'$ to $\widetilde{h}_1|_{\partial \widetilde{P}}$ and extend it over $\widetilde{P}$, keeping a neighbourhood of $\widetilde{h}_1^{-1}(M \times D)$ fixed, to an isotopy from $hd$ to a $PL$ homeomorphism $g : \widetilde{P} \longrightarrow M \times \mathbb{R}^k$. Then

$$\psi = g\widetilde{h}_1^{-1} : M \times \mathbb{R}^k \longrightarrow M \times \mathbb{R}^k$$

is the identity on $\partial M \times \mathbb{R}^k$ and on a neighbourhood of $M \times D$. The Alexander isotopy constructed for (4.1) slides $\psi$ to the identity, keeping $\partial M \times \mathbb{R}^k$ and a neighbourhood of $M \times D$ fixed. Therefore $g$ is isotopic to $\widetilde{h}_1$ leaving $\partial \widetilde{P}$ and a neighbourhood of $\widetilde{h}_1^{-1}(M \times D)$ fixed. Combining this isotopy with that from $hd$ to $g$ gives the required result.

**(5.2) Theorem.** *Let $P$ be a $PL$ manifold of dimension $m+r+1$ and $h : P \longrightarrow M \times T^r \times \mathbb{R}$ a proper homotopy equivalence of pairs. Then any splitting of $h|_{\partial P}$ extends to a splitting of $h$.*

**Proof.** Let $(N', s', \phi')$ be a splitting of $h|_{\partial P}$. Clearly $P$ has two ends, and $s'|_{N' \times [1,\infty)}$ provides a collar of those ends of $\partial P$ contained by one of the ends of $P$. Using the relative collaring theorem (3.2) we can extend this collar to a collar of the whole end. If $N$ denotes the base of the extended collar, then $\partial N = N'$. In exactly the same way we can produce a (disjoint) collar of the other end which is compatible with $s'|_{N' \times [-1,-\infty)}$. As in (4.4) we have an $h$-cobordism (this time between manifolds with boundary) sandwiched by the two collars. A version of Stallings [17] for manifolds with boundary provides a $PL$ homeomorphism $s : N \times \mathbb{R} \longrightarrow P$ such that $s|_{N' \times \mathbb{R}} = s'$. If we can find a proper homotopy $\phi$ between $hs$ and the usual product $\lambda \times \mathrm{id}_{\mathbb{R}}$ which extends $\phi'$, then $(N, s, \phi)$ is the required splitting of $h$. We can certainly extend $\phi'$ to a proper homotopy between $hs$ and some map $g : N \times \mathbb{R} \longrightarrow M \times T^r \times \mathbb{R}$. Then, proceeding as in (4.4), we can construct a proper homotopy from $g$ to $\lambda \times \mathrm{id}_{\mathbb{R}}$ which fixes $N' \times \mathbb{R}$. The composition of the two homotopies gives $\phi$.

A proof of our relative splitting theorem (2.2) may now be obtained simply by reworking the material of §4, allowing $M$ to have boundary and using (5.1) and (5.2) in place of (4.1) and (4.4).

We end this section with the observation that in the special case when $M$ and $h$ are both $PL$, we can extend the **natural splitting** of $h|_{\partial W}$ to a splitting of $h$.

## §6. The trivialization problem

This final section will be devoted to a proof of the following result.

**(6.1) Theorem.** *Let $X$ be a compact 4-connected polyhedron of dimension $m$, $\pi : E \longrightarrow X$ a $PL$ $\mathbb{R}^k$-bundle with $k \geq m+2$, and $t : E \longrightarrow X \times \mathbb{R}^k$ a topological trivialization. Then $t$ is properly homotopic to a $PL$ trivialization.*

As a direct corollary we have a solution to the bundle trivialization problem proposed in §1. A proof of (6.1) will therefore complete our arguments.

**Proof of (6.1).** Triangulate $X$ in some way, let $K$ denote the 4–skeleton of the triangulation, and $CK$ the cone on $K$. Since $X$ is 4–connected, the inclusion map of $K$ into $X$ extends to a map $d$ from all of $CK$ to $X$. Using $d$ we can pull back the diagram

$$\begin{array}{ccc} E & \xrightarrow{\ t\ } & X \times \mathbb{R}^k \\ \downarrow & & \downarrow \\ X & = & X \end{array}$$

to give a bundle over $CK$ and trivialization

$$\begin{array}{ccc} E^* & \xrightarrow{\ t^*\ } & CK \times \mathbb{R}^k \\ \downarrow & & \downarrow \\ CK & = & CK \end{array}$$

As $CK$ is contractible, we can find a $PL$ bundle equivalence

$$f \ : \ E^* \longrightarrow CK \times \mathbb{R}^k \ .$$

Now let $E(K)$ denote the part of $E$ which lies over $K$, and extend

$$K \times \mathbb{R}^k \xrightarrow{\ t^{-1}\ } E(K) \xrightarrow{\ f\ } K \times \mathbb{R}^k$$

to a map $g : X \times \mathbb{R}^k \longrightarrow X \times \mathbb{R}^k$ as follows. Use the contractibility of $CK$ again to produce a map $r : X \longrightarrow CK$ which extends the identity on $K$, and define

$$g(x, u) \ = \ (x, p_2 f t^{*-1}(r(x), u)) \ .$$

Then $g$ is a bundle equivalence and is homotopic to the identity via a proper homotopy. Therefore $gt : E \longrightarrow X \times \mathbb{R}^k$ is a topological trivialization of our bundle which is properly homotopic to $t$ and which, by construction, is $PL$ over $K$.

We now apply the splitting process over each simplex $\Delta$ of $X$, in other words we split $t : E(\Delta) \longrightarrow \Delta \times \mathbb{R}^k$, taking care that the splittings fit together to give a splitting of $t : E \longrightarrow X \times \mathbb{R}^k$. Since $t$ is already $PL$ over the 4–skeleton of $X$, we may use the natural splitting over each simplex of $K$. These splittings are of course compatible, in the sense that the natural splitting over a simplex restricts to the natural splitting over any face. Having rid ourselves of low dimensional problems in this way, we work on the remaining simplexes inductively in order of increasing dimension. The relative splitting theorem allows us to construct a Novikov splitting over each simplex which, when restricted to a face is the splitting constructed earlier. Suppose $(B, s, \phi)$ is the splitting over $\Delta$. We observe that $B$ is $PL$ homeomorphic to $\Delta$. If $\Delta \in K$, then $B = \Delta$, and for the other simplexes we can use the Poincaré Conjecture noting that, in the special case of a 5-simplex, we know the boundary is already standard. Therefore we have a compatible system of $PL$ homeomorphisms

$$s \;:\; \Delta \times \mathbb{R}^k \longrightarrow E(\Delta) \;\; (\Delta \in X)$$

and a compatible family of proper homotopies $\phi : \Delta \times \mathbb{R}^k \longrightarrow \Delta \times \mathbb{R}^k$ from $ts$ to $\lambda \times \mathrm{id}_{\mathbb{R}^k}$.

A homeomorphism from a ball to itself, which is the identity on the boundary, is isotopic to the identity keeping the boundary fixed. Therefore, again taking the simplexes in some order of increasing dimension, we can inductively homotope the $\lambda$'s to the identity. Combining all these homeomorphisms and homotopies gives a $PL$ homeomorphism

$$s \;:\; X \times \mathbb{R}^k \longrightarrow E$$

together with a proper homotopy from $ts$ to the identity. Hence $t$ is homotopic to $s^{-1}$ via a proper homotopy that is fixed over $K$. Although $s^{-1}$ sends $\Delta \times \mathbb{R}^k$ to $E(\Delta)$, for each $\Delta \in X$, it is not at this stage a bundle map. If $\Gamma$ is a $PL$ section of $E \xrightarrow{\pi} X$ there is an ambient isotopy $H$ of $E$ such that $H_1 s^{-1}(\Delta \times \{0\}) = \Gamma(\Delta)$ for every simplex $\Delta$ of $X$. To construct $H$ we use the Unknotting Theorem [21] inductively. For the inductive step we have a situation where $s^{-1}(\Delta \times \{0\})$ and $\Gamma(\Delta)$ are two embeddings of $\Delta$ into $E(\Delta)$ which agree on $\partial\Delta$, and which are therefore ambient isotopic keeping $\partial\Delta$ fixed. The section $\Gamma$ now has two normal bundles in $E$, namely the bundle structure of $E$ itself, and that given by $s^{-1}$. The stable range uniqueness theorem for $PL$ normal bundles [7] provides a $PL$ ambient isotopy $G$ of $E$ such that

$$G_1 H_1 s^{-1} \;:\; X \times \mathbb{R}^k \longrightarrow E$$

is fibre preserving. Therefore $G_1 H_1 s^{-1}$ is a $PL$ trivialization of $E \xrightarrow{\pi} X$ which is properly homotopic to $t$, and the proof of (6.1) is complete.

## References

[1] H. Bass, A. Heller and R. G. Swan, **The Whitehead group of a polynomial extension**, Publ. Math. I.H.E.S. 22 (1964) 61–79.

[2] W. Browder, **Homotopy type of differentiable manifolds**, Proceedings of the Aarhus Symposium, 1962, 42–46.
reprinted in Proceedings of 1993 Oberwolfach Conference on the Novikov Conjectures, Index Theorems and Rigidity (ed. S.Ferry, A.Ranicki and J.Rosen

berg), Vol. 1, LMS Lecture Notes 226, Cambridge (1995) 97–100.

[3] W. Browder, **Surgery on simply connected manifolds**, Ergebnisse der Mathematik 65, Springer Verlag (1972).

[4] W. Browder and M. W. Hirsch, **Surgery on piecewise linear manifolds and applications**, Bull. Amer. Math. Soc. 72 (1966) 959–964.

[5] D. B. A. Epstein, **Ends**, Topology of 3-manifolds, Proceedings of the 1961 Georgia Symposium, Prentice Hall (1962) 110–117.

[6] H. Gluck, **Embeddings in the trivial range**, Annals of Math. 81, (1965) 195–210.

[7] A. Haefliger and C. T. C. Wall, **Piecewise linear bundles in the stable range**, Topology 4 (1965) 209–214.

[8] M. W. Hirsch, **On normal microbundles**, Topology 5 (1966) 229–240.

[9] J. M. Kister, **Microbundles are fibre bundles**, Bull. Amer. Math. Soc. 69 (1963) 854–857.

[10] R. K. Lashof and M. G. Rothenberg, **On the Hauptvermutung, triangulation of manifolds and $h$-cobordism**, Bull. Amer. Math. Soc. 72 (1966) 1040–1043.

[11] R. K. Lashof and M. G. Rothenberg, **Hauptvermutung for manifolds**, Conference on the topology of manifolds, Prindle, Weber and Schmidt (1968) 81–105.

[12] J. Milnor, **Lectures on the $h$-cobordism theorem**, Princeton (1965).

[13] S. P. Novikov, **Topological invariance of rational Pontrjagin classes**, Doklady 163 (1965) 298–300. (Soviet Math. (1965) 921–923.)

[14] S. P. Novikov, **Homotopically equivalent smooth manifolds I**, Izv. Akda. Nauk. S.S.S.R. Ser. Mat. 28 (1964) 365–474. (Translations Amer. Math. Soc. 48 (1965) 271–396.)

[15] C. P. Rourke and B. J. Sanderson, **Piecewise linear topology**, Ergebnisse der Mathematik 69, Springer (1972).

[16] L. C. Siebenmann, **The obstruction to finding a boundary for an open manifold of dimension greater than five**, Ph.D. Thesis, Princeton (1966).

[17] J. R. Stallings, **On infinite processes leading to differentiability in the complement of a point**, Differential and Combinatorial Topology, Princeton (1965) 245–254.

[18] D. P. Sullivan, **On the Hauptvermutung for manifolds**, Bull. Amer. Math. Soc. 73 (1967) 598–600.

[19] D. P. Sullivan, **Triangulating and smoothing homotopy equivalences and homeomorphisms**, Geometric Topology Seminar Notes, Princeton (1967). (Published in this volume).

[20] R. E. Williamson Jr., **Cobordism of combinatorial manifolds**, Annals of Math. 83 (1966) 1–33.

[21] E. C. Zeeman, **Unknotting combinatorial balls**, Annals of Math. 78 (1963) 501–526.

[22] E. C. Zeeman, **Relative simplicial approximation**, Proc. Camb. Phil. Soc. 60 (1964) 39–43.

Institute for Advanced Study
Princeton
Spring 1968

# CHAPTER II

## The Hauptvermutung according to Casson and Sullivan

**By C. P. Rourke**

### §1. Introduction

Suppose $M$ and $Q$ are $PL$ manifolds and $h : Q \longrightarrow M$ is a (topological) homeomorphism. The Hauptvermutung asserts that in this situation there is a $PL$ isomorphism $g : Q \longrightarrow M$. The purpose of Chapter II is to give a proof of the Hauptvermutung for a large class of manifolds:

**(1.1) Main theorem.** *Let $M, Q, h$ be as above. Suppose:*
(1) *$M$ and $Q$ are closed of dimension $\geq 5$ or bounded of dimension $\geq 6$.*
(2) *$\pi_1(M) = \pi_1(\partial M) = 0$. (Assumed for each component of $M$ and $\partial M$.)*
(3) *$H^3(M; \mathbb{Z}_2) = 0$.*
*Then $h$ is homotopic to a $PL$ isomorphism.*

We shall also indicate a proof that condition (3) can be weakened to the following:

(3′) *$H^4(M; \mathbb{Z})$ has no elements of order 2.*

Various refinements of the theorem are possible. One can weaken condition (2) to $\pi_1(M) = \pi_1(\partial M)$ if $\partial M$ is connected and non-empty and $h$ is a simple homotopy equivalence. One can keep submanifolds, on which $h$ is already a $PL$ isomorphism, fixed during the homotopy. Precise statements of these refinements are given in §7.

The rest of the introduction consists of a broad outline of the proof of the main theorem followed by a guide to the rest of the chapter.

**Outline of the Proof.** According to Sullivan [41], $h$ is homotopic to a $PL$ isomorphism if and only if a certain map $q_h : M_0 \longrightarrow G/PL$ is null-homotopic (where $M_0 = M$ if $\partial M \neq \emptyset$ and $M \backslash \{\text{pt.}\}$ if $\partial M = \emptyset$) and from the definition of $q_h$ if follows that $q_h$ factors via $TOP/PL$.

$$\begin{array}{ccc} & TOP/PL & \\ \nearrow & & \searrow^{i} \\ M_0 & \xrightarrow{\quad q_h \quad} & G/PL \end{array}$$

The spaces $PL$, $TOP$ and $G$ will be defined in §2 and an account of the result from

[41] which we use is given in §3, where a more general result is proved. The main tools here are Browder-Novikov-Wall style surgery and the $h$-cobordism theorem.

From the above diagram we can assert that $q_h$ is null-homotopic if we know that $i$ is null-homotopic. In fact we prove that $i$ factors via $K(\mathbb{Z}_2, 3)$

$$\begin{array}{ccccc} & & K(\mathbb{Z}_2,3) & & \\ & \nearrow & & \searrow & \\ TOP/PL & & \xrightarrow{\quad i \quad} & & G/PL \end{array}$$

From this factoring it follows that the obstruction to homotoping $q_h$ to zero is an element of $H^3(M_0; \mathbb{Z}_2)$ and, using condition (3), the result follows.

The proof of the factorization of $i$ has two main steps:

**Step 1.** Construct a periodicity map

$$\mu \; : \; G/PL \longrightarrow \Omega^{4n}(G/PL)$$

where $\Omega^n(X)$ denotes the $n$-th loop space on $X$. $\mu$ is defined to be the composite

$$G/PL \xrightarrow{\alpha} (G/PL)^{\mathbb{CP}^{2n}} \xrightarrow{\sigma} \Omega^{4n}(G/PL)$$

where $X^Y$ is the space of maps $Y \longrightarrow X$, and $\alpha$ is defined by $\alpha x(y) = x$ for all $y \in \mathbb{CP}^{2n}$ (complex projective $2n$-space) and $\sigma$ is a canonically defined surgery obstruction (see §4). The periodicity map $\mu$ has the property that $\mu_* : \pi_k(G/PL) \longrightarrow \pi_{k+4n}(G/PL)$ is an isomorphism for $k \neq 4$ and multiplies by 2 for $k = 4$. (Recall that $\pi_n(G/PL) = 0$ for $n$ odd, $\mathbb{Z}$ if $n = 4k$ and $\mathbb{Z}_2$ if $n = 4k+2$, essentially Kervaire and Milnor [19], see §4). The proof that $\mu_*$ has these properties follows from a product formula for surgery obstructions which is proved by Rourke and Sullivan [36]. It follows that the homotopy-theoretic fibre of $\mu$ is a $K(\mathbb{Z}_2, 3)$.

**Step 2.** Prove that the composite

$$\sigma' \; : \; (TOP/PL)^{\mathbb{CP}^{2n}} \longrightarrow (G/PL)^{\mathbb{CP}^{2n}} \xrightarrow{\sigma} \Omega^{4n}(G/PL)$$

is null-homotopic. This result can be elucidated as follows: the surgery obstruction for a map $\mathbb{CP}^{2n} \longrightarrow G/PL$ is an "index" obstruction (see §4) and can be measured in terms of the Pontrjagin numbers of a certain bundle over $\mathbb{CP}^{2n}$. Using the fact that the map comes from a map of $\mathbb{CP}^{2n} \longrightarrow TOP/PL$, it follows from Novikov [27] that the obstruction is zero. Thus to prove the result it is only necessary to prove Novikov's result in a canonical form, and this is done by using Siebenmann's thesis [37]. In fact we never mention Pontrjagin classes but prove the result directly using Siebenmann. Details are in §5.

Now consider the diagram

$$\begin{array}{ccc}
TOP/PL & \xrightarrow{\quad i \quad} & G/PL \\
{\scriptstyle \alpha'}\Big\downarrow & & \Big\downarrow{\scriptstyle \alpha} \\
(TOP/PL)^{\mathbb{CP}^{2n}} & \longrightarrow & (G/PL)^{\mathbb{CP}^{2n}} \\
& {\scriptstyle \sigma'}\searrow & \Big\downarrow{\scriptstyle \sigma} \\
& & \Omega^{4n}(G/PL)
\end{array}$$

$\sigma'\alpha' \simeq *$, consequently $i$ factors via the fibre of $\sigma\alpha$ which is $K(\mathbb{Z}_2, 3)$, as required.

### Guide to the rest of Chapter II

§2 collects most of the notation and basic definitions which we use. Two important definitions here are the semi-simplicial complexes $HT(M)$ and $NM(M)$. These are (roughly) the space of homotopy triangulations of $M$ and the space of "normal maps" onto $M$. A normal map is a degree 1 map $f : M_1 \longrightarrow M$ covered by a bundle map from the normal bundle of $M_1$ to some bundle over $M$ (the terminology "normal map" is Browder's). $HT(M)$ should not be confused with the **set** of $PL$ equivalence classes of homotopy triangulations of $M$, which we denote $Ht(M)$. (This set was called $PL(M)$ by Sullivan [41] – a notation which we consider should be reserved for the space of $PL$ isomorphisms of $M$.)

In §3 we prove two basic homotopy equivalences:

$$NM(M) \simeq (G/PL)^M$$

(which is true in general) and

$$HT(M) \simeq NM(M)$$

if $M$ is bounded of dimension $\geq 6$ and $\pi_1(M) = \pi_1(\partial M) = 0$. The classification of homotopy triangulations

$$Ht(M) \cong [M_0, G/PL]$$

follows at once. We conclude §3 by defining the characteristic map $q_h : M_0 \longrightarrow G/PL$ (see sketch above) and proving that it factors via $TOP/PL$ if $h$ is a homeomorphism.

The main result of §4 is the homotopy equivalence

$$(G/PL)^M \simeq (G/PL)^{M_0} \times \Omega^n(G/PL)$$

in case $M$ is closed of dimension $\geq 6$. The "canonical surgery obstruction" map

$$\sigma : (G/PL)^M \longrightarrow \Omega^n(G/PL)$$

is then projection on the second factor. We deduce the properties of $\mu$ mentioned earlier.

§5 completes the proof of proving Step 2 above. We start from a refined version of the Novikov–Siebenmann splitting theorem and construct a map

$$\lambda : (TOP/PL)^M \longrightarrow HT(M)$$

which commutes with maps of both spaces into $(G/PL)^M$ – the first induced by inclusion, the second defined in §3. The result then follows easily from the definition of $\sigma$.

In the remaining sections, we consider improvements to the main theorem. In §6 we sketch the proof of the weakening of condition (3).

In §7 three refinements are proved:

(a) Replace condition (2) (simple connectivity) by the condition that $\pi_1(\partial M) \longrightarrow \pi_1(M)$ is an isomorphism **and** $h$ is a simple homotopy equivalence. A corollary (using Connell and Hollingsworth [7]) is that the Hauptvermutung holds for manifolds with 2-dimensional spines.

(b) Assume, instead of a topological homeomorphism $h : Q \longrightarrow M$, that $M$ and $Q$ are (topologically) $h$-cobordant.

(c) Assume that $h : Q \longrightarrow M$ is a cell-like map (cf. Lacher [21]) rather than a homeomorphism.

In §8 we prove a theorem on homotopy triangulations of a block bundle. Two corollaries are:

1. A relative Hauptvermutung; that is, if $h : (M_1, Q_1) \longrightarrow (M, Q)$ is a homeomorphism of pairs and $Q$ is a submanifold either of codim 0 or codim $\geq 3$, then $h$ is homotopic to a $PL$ isomorphism of pairs.
2. The embedding theorem first proved by Casson-Sullivan and independently by Haefliger [11] and Wall.

I am indebted to Chris Lacher for pointing out §7(b) and a crucial step in §7(c), and to Greg Brumfiel and George Cooke for patiently explaining §6 . Chapter III by Cooke supplies more detail for §6.

## §2. Notation and basic definitions

We refer to Rourke and Sanderson [35] for the definition of the $PL$ category. Objects and maps in this category will be prefixed "$PL$".

The following are standard objects in the category. $\mathbb{R}^n = \mathbb{R}^1 \times \cdots \times \mathbb{R}^1$, Euclidean $n$-space. $\Delta^n$, vertices $\{v_0, \cdots, v_n\}$, the standard $n$-simplex. The face map $\partial_i^n : \Delta^{n-1} \longrightarrow \Delta^n$ is the simplicial embedding which preserves order and fails to cover $v_i$ and $\Lambda_{n,i} = c\ell(\partial\Delta^n \backslash \partial_i\Delta^{n-1})$. The $n$-cube $I^n = [-1,+1]^n$ and $I = [0,1]$, the unit interval.

**Semi-simplicial objects.**

We work always without degeneracies – by Rourke and Sanderson [33] they are irrelevant to our purposes, and we shall not then have to make arbitrary choices to define them.

Let $\Delta$ denote the category whose objects are $\Delta^n$, $n = 0, 1, \ldots$ and morphisms generated by $\partial_i^n$. A **$\Delta$-set, pointed $\Delta$-set, $\Delta$-group** is a contravariant functor from $\Delta$ to the category of sets, pointed sets, groups.

$\Delta$-maps, $\Delta$-homomorphisms, etc. are natural transformations of functors. Ordered simplicial complexes are regarded as $\Delta$ sets in the obvious way. A $\Delta$-set $X$ satisfies the **(Kan) extension condition** if every $\Delta$-map $\Lambda_{(n,i)} \longrightarrow X$ possesses an extension to $\Delta^n$.

We now define the various $\Delta$-groups and $\Delta$-monoids that we need:

$PL_q$ : typical $k$-simplex is a zero and fibre preserving $PL$ isomorphism

$$\sigma : \Delta^k \times \mathbb{R}^q \longrightarrow \Delta^k \times \mathbb{R}^q$$

(i.e $\sigma|_{\Delta^k \times \{0\}} =$ id. and $\sigma$ commutes with projection on $\mathbb{R}^q$).

$\widetilde{PL}_q$ : typical $k$-simplex is a zero and block preserving $PL$ isomorphism

$$\sigma : \Delta^k \times \mathbb{R}^q \longrightarrow \Delta^k \times \mathbb{R}^q$$

(i.e. $\sigma|_{\Delta^k \times \{0\}} =$ id and $\sigma(K \times \mathbb{R}^q) = K \times \mathbb{R}^q$ for each subcomplex $K \subset \Delta^k$).

Face maps are defined by restriction and $PL_q, \widetilde{PL}_q$ form $\Delta$-groups by composition.

$G_q$ : typical $k$-simplex is a zero and fibre preserving homotopy equivalence of pairs

$$\sigma : (\Delta^k \times \mathbb{R}^q, \Delta^k \times \{0\}) \longrightarrow (\Delta^k \times \mathbb{R}^q, \Delta^k \times \{0\})$$

(i.e. $\sigma^{-1}(\Delta^k \times \{0\}) = \Delta^k \times \{0\}$ and $\sigma|_{\Delta^k \times (\mathbb{R}^q \backslash \{0\})}$ has degree $\pm 1$).

$\widetilde{G}_q$ : typical $k$-simplex is a zero and block preserving homotopy equivalence of pairs

$$\sigma : (\Delta^k \times \mathbb{R}^q, \Delta^k \times \{0\}) \longrightarrow (\Delta^k \times \mathbb{R}^q, \Delta^k \times \{0\}) .$$

Again face maps are defined by restriction and $G_q$, $\widetilde{G}_q$ form $\Delta$-monoids by composition.

Inclusions $i : PL_q \subset PL_{q+1}$ etc. are defined by $i(\sigma) = \sigma \times$ id. (write $\mathbb{R}^{q+1} = \mathbb{R}^q \times \mathbb{R}^1$) and the direct limits are denoted $PL, \widetilde{PL}, G, \widetilde{G}$.

The notation used here differs from that used in Rourke and Sanderson [32], where these complexes were called $PL_q(\mathbb{R})$ etc. However, as we never use the other complexes, no confusion should arise.

$G/PL_q$ and $\widetilde{G}_q/\widetilde{PL}_q$ are the complexes of right cosets (i.e. a $k$-simplex of $G_q/PL_q$ is an equivalence class of $k$-simplexes of $G_q$ under $\sigma_1 \sim \sigma_2$ iff $\sigma_1 = \sigma_3 \circ \sigma_2$ where $\sigma_3 \in PL_q$).

The following basic properties of the complexes defined so far will be used (cf. Rourke and Sanderson [33] for notions of homotopy equivalence, etc.).

**(2.1) Proposition.**

(a) *$G_q \subset \widetilde{G}_q$ is a homotopy equivalence for all $q$.*

(b) *The inclusions $PL_q \subset \widetilde{PL}_q$, $PL_q \subset PL$, $\widetilde{PL}_q \subset \widetilde{PL}$ and $G_q \subset G$ are all $(q-1)$-connected.*

(c) *$PL \subset \widetilde{PL}$ is a homotopy equivalence.*

(d) *The map $\widetilde{G}_q/\widetilde{PL}_q \longrightarrow \widetilde{G}/\widetilde{PL}$ induced by inclusion is a homotopy equivalence for $q > 2$.*

(e) *The complexes $G_q/PL_q$ (resp. $\widetilde{G}_q/\widetilde{PL}_q$) are classifying for $PL$ bundles with fibre $(\mathbb{R}^q, \{0\})$ and with a fibre homotopy trivialization (resp. open block bundles with a block homotopy trivialization – i.e. a trivialization of the associated fibre space).*

**Remark.** A "$PL$ bundle with a fibre homotopy trivialization" means a pair $(\xi^q, h)$ where $\xi^q/K$ is a $PL$ fibre bundle with base $K$ and fibre $(\mathbb{R}^q, \{0\})$, and

$$h : E_0(\xi^q) = E(\xi)\backslash K \longrightarrow K \times (\mathbb{R}^q\backslash\{0\})$$

is a fibre map with degree $\pm 1$ on each fibre (cf. Dold [9]). Such pairs form a bundle theory with the obvious definitions of induced bundle, Whitney sum, etc. (see Sullivan [41]). A $PL$ block bundle with a block homotopy trivialization can be defined in a similar way.

**Proof of 2.1.** Parts (a) to (d) were all proved in Rourke and Sanderson [32], the following notes will help the reader understand the status of these results:

(a) is proved by an easy "straight line" homotopy.

(b) the first two parts depend on Haefliger and Poenaru [13] – the second part is explicit in Haefliger and Wall [14] and the first part is a translation of their main result. The third part is a straight analogue of the smooth stability theorem and the fourth part is classical homotopy theory (James [17]).

(c) follows from (b).

(d) is a translation of the stability theorem of Levine [23] using the transverse regularity of Rourke and Sanderson [31] and Williamson [45] – special arguments are necessary in low dimension – see §7.

(e) follows from the fibrations

$$G_q/PL_q \longrightarrow BPL_q \longrightarrow BG_q \ ,$$
$$\widetilde{G}_q/\widetilde{PL}_q \longrightarrow B\widetilde{PL}_q \longrightarrow B\widetilde{G}_q$$

(see Rourke and Sanderson [34]) and can also be proved by a simple direct argument analogous to Rourke and Sanderson [30; §5].

$TOP_q$ is the topological analogue of $PL_q$ i.e. a $k$-simplex is a zero and fibre preserving homeomorphism $\Delta^k \times \mathbb{R}^q \longrightarrow \Delta^k \times \mathbb{R}^q$. $TOP$ is the direct limit of

$$i \ : \ TOP_q \subset TOP_{q+1} \subset \ldots \ .$$

The stability theorem for $TOP_q$ is for weaker than 2.1(b), but one can define a stable $K$-theory of topological bundles (see Milnor [26]).

**(2.2) Proposition.** *The complex $TOP_q/PL_q$ classifies PL bundles with a topological trivialization.*

The proof is the same as 2.1(e).

All the complexes defined so far satisfy the extension condition – this follows easily from the existence of a $PL$ isomorphism $\Lambda_{k,i} \times I \longrightarrow \Delta^k$.

**Function spaces.**

Let $X$ be a $\Delta$-set with the extension condition and $P$ a polyhedron. A **map** of $P$ in $X$ is an ordered triangulation $K$ of $P$ and a $\Delta$-map $K \longrightarrow X$. A typical $k$-simplex of the $\Delta$-set $X^P$ is a map $P \times \Delta^k \longrightarrow X$ where the triangulation $K$ of $P \times \Delta^k$ contains $P \times \partial_i \Delta^{k-1}$ as a subcomplex, each $i$. Face maps are then defined by restriction. For connections with other definitions see Rourke and Sanderson [33].

When $X$ is pointed, denote by $* \subset X$ the subset consisting of base simplexes (or the identity simplexes in case $X$ is a $\Delta$-group or monoid). $X^P$ is then pointed in the obvious way.

Relative function spaces are defined in a similar way. In particular the $n$th loop space of $X$ is defined by

$$\Omega^n(X) \ = \ (X, *)^{(I^n, \partial I^n)} \ .$$

**$(\Delta^k, n)$-manifolds.**

Let $\mathcal{M}$ denote the category with objects: $PL$ manifolds; morphisms: inclusions of one $PL$ manifold in the boundary of another. A **$(\Delta^k, n)$-manifold** is a lattice in $\mathcal{M}$ isomorphic to (and indexed by) the lattice of faces of $\Delta^k$, where the isomorphism is graded and decreases dimension by $n$. (I.e. each $s$-face of $\Delta^k$ indexes an $(n+s)$-manifold.) If $M^{n,k}$ is a $(\Delta^k, n)$-manifold then the element indexed by $\sigma \in \Delta^k$ is denoted $M_\sigma$.

**Examples.**

(1) If $M^n$ is an $n$-manifold then $M^n \times \Delta^k$ is a $(\Delta^k, n)$-manifold in the obvious way, with $M_\sigma = M \times \sigma$.

(2) A $(\Delta^0, n)$-manifold is an $n$-manifold.

(3) A $(\Delta^1, n)$-manifold is a cobordism of $n$-manifolds, possibly with boundary.

(4) If $M^{n,k}$ is a $(\Delta^k, n)$-manifold then the $(\Delta^k, n-1)$-manifold $\partial M^{n,k}$ is defined by

$$\begin{aligned}(\partial M)_{\Delta^k} &= c\ell\Big(\partial(M_{\Delta^k})\backslash \bigcup_i M_{\partial_i \Delta^{k-1}}\Big)\ , \\ (\partial M)_\sigma &= (\partial M)_{\Delta^k} \cap M_\sigma\ .\end{aligned}$$

Thus, in example (3), $\partial M^{n,1}$ is the corresponding cobordism between the boundaries.

Now we come to two basic definitions:

**The $\Delta$-set $HT(M)$.**

A **map** of $(\Delta^k, n)$-manifolds $f : M^{n,k} \longrightarrow Q^{n,k}$ is a map $f : M_{\Delta^k} \longrightarrow Q_{\Delta^k}$ such that $f(M_\sigma) \subset Q_\sigma$ for each $\sigma \in \Delta^k$. A **homotopy equivalence** of $(\Delta^k, n)$-manifolds is a map $h$ such that $h|_{M_\sigma} : M_\sigma \longrightarrow Q_\sigma$ is a homotopy equivalence for each $\sigma \in \Delta^k$.

Let $M^n$ be a $PL$ manifold (possibly with boundary). A $k$-simplex of the $\Delta$-set $HT(M)$ ("homotopy triangulations of $M$") is a homotopy equivalence of pairs

$$h : (Q^{n,k}, \partial Q^{n,k}) \longrightarrow (M^n \times \Delta^k, \partial M^n \times \Delta^k)$$

where $Q^{n,k}$ is some $(\Delta^k, n)$-manifold. (I.e. $h(\partial Q)_{\Delta^k} \subset \partial M^n \times \Delta^k$ and $h|_{\partial Q^{n,k}}$ is also a homotopy equivalence.)

Face maps are defined by restriction and it is easy to prove that $HT(M)$ satisfies the extension condition.

**The $\Delta$-set $NM(M)$.**

A typical $k$-simplex is a **normal map** $f : Q^{n,k} \longrightarrow M \times \Delta^k$. I.e. $f$ is a $(\Delta^k, n)$-map, has degree 1 on each pair $(Q_\sigma, \partial Q_\sigma) \longrightarrow (M \times \sigma, \partial M \times \sigma)$, and is covered by a map of $PL$ bundles:

$$\begin{array}{ccc} E(\nu_Q) & \xrightarrow{\widehat{f}} & E(\xi) \\ \downarrow & & \downarrow \\ Q & \xrightarrow{f} & M \times \Delta^k \end{array}$$

where $\nu_Q$ is the (stable) normal bundle of $Q$ and $\xi$ is some $PL$ bundle on $M \times \Delta^k$. Face maps are defined by restriction and it is again easy to check that $NM(M)$ satisfies the extension condition.

The (stable) **normal bundle** of $Q^{n,k}$ is the normal bundle of an embedding $Q^{n,k} \subset I^{n+N} \times \Delta^k$ ($N$ large) of $(\Delta^k, n)$-manifolds. The normal bundle of $Q^{n,k}$ restricts to the normal bundle of $Q_\sigma \subseteq I^{n+N} \times \sigma$ for each $\sigma$. To find such a bundle, use general position to embed, and apply Haefliger and Wall [14].

**Reducibility.**

Let $\xi^N / Q^{n,k}$ be a bundle. $T(\xi)$, the Thom space, is said to be **reducible** if there is a map $f : I^{n+N} \times \Delta^k \longrightarrow T(\xi)$ which respects the lattice structure and such that

$$f| \ : \ (I^{n+N} \times \sigma, \partial(I^{n+N} \times \sigma)) \longrightarrow (T(\xi|M_\sigma), T(\xi|\partial M_\sigma))$$

has degree 1 for each $\sigma$. Thus $f$ gives a simultaneous reduction of all the Thom spaces in the lattice.

For example, the Thom construction shows $T(\nu_Q)$ is reducible, in fact has a canonical choice of reduction map.

Notice that, in the definition of $NM(M), T(\xi)$ is reducible. This is because the Thom isomorphism is natural and $f$ has degree 1. Indeed $T(\widehat{f})$ and the canonical reduction of $T(\nu_Q)$ give a reduction of $T(\xi)$.

**The $\Delta$-sets $(G/PL)_M$ and $(TOP/PL)_M$.**

Finally we define two $\Delta$-sets which, although essentially the same as the function spaces $(G/PL)^M$ and $(TOP/PL)^M$, have certain advantages for some of our constructions.

A $k$-simplex of the $\Delta$-set $(G/PL)_M$ (resp. $(TOP/PL)_M$) is a stable $PL$ bundle $\xi^N/M \times \Delta^k$ together with a fibre homotopy trivialization (resp. a topological trivialization). Face maps are defined by restriction and the extension condition is easily verified.

**(2.3) Proposition.** *There are homotopy equivalences*

$$\kappa : (G/PL)_M \longrightarrow (G/PL)^M ,$$
$$\kappa' : (TOP/PL)_M \longrightarrow (TOP/PL)^M$$

*which commute with the natural maps*

$$(TOP/PL)^M \xrightarrow{i_*} (G/PL)^M \quad , \quad (TOP/PL)_M \xrightarrow{j} (G/PL)_M .$$

This follows at once from 2.1(e) and 2.2, commutativity being obvious. One of the advantages of the new sets is that they have an easily described $H$-space structure. For example, the map

$$m : (G/PL)_M \times (G/PL)_M \longrightarrow (G/PL)_M ;$$
$$(\xi_1^{N_1}, t_1; \xi_2^{N_2}, t_2) \longmapsto (\xi_1^{N_1} \oplus \xi_2^{N_2}, t_1 \oplus t_2)$$

endows $(G/PL)_M$ with a homotopy commutative $H$-space structure with homotopy unit. [To make precise sense of $t_1 \oplus t_2$, when $\xi_1$ and $\xi_2$ are stable bundles, regard the range of each as $M \times \Delta^k \times \mathbb{R}^\infty$ and choose a homeomorphism $\mathbb{R}^\infty \times \mathbb{R}^\infty \longrightarrow \mathbb{R}^\infty$ by alternating coordinates.]

"$\times$", in the displayed formula above, means categorical direct product (as always when we are dealing with $\Delta$-sets). For sets with the extension condition $X \times Y$ has the same homotopy type as $X \otimes Y$ (cf. Rourke and Sanderson [33]).

## §3. An account of Sullivan theory

In this section we prove the following results:

**(3.1) Theorem.** *If $M^n$ is any PL manifold (with or without boundary) then there is a homotopy equivalence*

$$r : NM(M^n) \longrightarrow (G/PL)^{M^n} .$$

**(3.2) Theorem.** *If $M^n$, $n \geq 6$ is a PL manifold with non-empty boundary, and $\pi_1(M) = \pi_1(\partial M) = 0$, then there is a homotopy equivalence*

$$i : HT(M) \longrightarrow NM(M) .$$

Theorem 3.1 is proved by transverse regularity (the usual argument) while Theorem 3.2 is proved by surgery. This breaks the usual Browder-Novikov argument into its two basic components. Combining the two results we have:

**(3.3) Corollary.** *If $M$ is as in Theorem 3.2 then there is a homotopy equivalence*

$$q : HT(M) \longrightarrow (G/PL)^M .$$

To obtain a theorem for closed manifolds, suppose $M$ is closed and let $M_0 = M\backslash\{disc\}$, with $n \geq 6$. Define a $\Delta$-map

$$c : HT(M_0) \longrightarrow HT(M)$$

by adding a conc over the boundary. (The Poincaré theorem is used here).

**(3.4) Theorem.** *$c$ is a homotopy equivalence.*

**(3.5) Corollary.** *If $M$ is closed of dimension $\geq 6$ and $\pi_1(M) = 0$, then there is a homotopy equivalence $HT(M) \longrightarrow (G/PL)^{M_0}$.*

**Classification of homotopy equivalences.**

Now let $M^n$ ($n \geq 5$) be a $PL$ manifold, $\pi_1(M) = \pi_1(\partial M) = 0$ (and if $n = 5$ assume $\partial M = \emptyset$). Define the set $Ht(M)$ (homotopy triangulations of $M$) as follows:

A representative is a homotopy equivalence $h : (Q^n, \partial Q^n) \longrightarrow (M^n, \partial M^n)$ where $Q^n$ is some $PL$ manifold. $h_1 \sim h_2$ if there is a $PL$ isomorphism $g : Q_1 \longrightarrow Q_2$ such that

$$\begin{array}{ccc} Q_1 & \xrightarrow{\;g\;} & Q_2 \\ & {\scriptstyle h_1}\searrow \quad \swarrow{\scriptstyle h_2} & \\ & M & \end{array}$$

is homotopy commutative. If follows immediately from the $h$-cobordism theorem that $Ht(M) = \pi_0(HT(M))$.

**(3.6) Corollary.** *Let $n \geq 6$ and write $M_0^n = M^n$ if $\partial M \neq \emptyset$ and $M_0^n = M\backslash\{disc\}$ if $\partial M = \emptyset$. ($\pi_1(M) = \pi_1(\partial M) = 0$, as usual.) Then there is a bijection*

$$q_* : Ht(M) \longrightarrow [M_0, G/PL]$$

*where $[\ ,\ ]$ denotes homotopy classes.*

This follows immediately from 3.3 and 3.4. For the case $n = 5$ of the main theorem we need

**(3.7) Addendum.** *If $n = 5$ and $\partial M = \emptyset$ then there is an injection*

$$q_* : Ht(M) \longrightarrow [M, G/PL] .$$

This will be proved using methods similar to those used for 3.2 - 3.4. Following these proofs, we shall make two remarks about $q_*$ which clarify the properties needed for the main theorem.

**Proof of Theorem 3.1.** We define homotopy inverses $r_1 : NM(M) \longrightarrow (G/PL)_M$ and $-r_2 : (G/PL)_M \longrightarrow NM(M)$. Then $r = \kappa \circ r_1$ is the required equivalence.

**Definition of $r_1$.**

**0-simplexes.**

Let $f : Q^n \longrightarrow M^n$ be a degree 1 map covered by a bundle map:

$$\begin{array}{ccc} E(\nu_Q) & \xrightarrow{\hat{f}} & E(\xi) \\ \downarrow & & \downarrow \\ Q^{n,k} & \xrightarrow{f} & M \end{array}$$

Then, as remarked earlier, $T(\xi)$ has a prescribed reduction. Let

$$u \;:\; (I^{n+N}, \partial I^{n+N}) \longrightarrow (T(\xi), T(\xi|\partial M))$$

be this reduction. Then the uniqueness theorem of Spivak [38] (see the proof given in Wall [43]) says that there is a fibre homotopy equivalence $g : \xi \longrightarrow \nu_M$ such that

(3.8)
$$\begin{array}{ccc} & (I^{n+N}, \partial I^{n+N}) & \\ {\scriptstyle u}\swarrow & & \searrow{\scriptstyle u'} \\ (T(\xi), T(\xi|_{\partial M})) & \xrightarrow{\quad T(g) \quad} & (T(\nu_M), T(\nu_{\partial M})) \end{array}$$

is homotopy commutative, where $u'$ is the canonical reduction of $\nu_M$. This diagram determines $g$ up to fibre homotopy, see Wall [43].

Now $g$ determines a stable fibre homotopy trivialization of $[\nu_M] - [\xi]$ (where $[\xi]$ denotes the element of the $K$-theory corresponding to $\xi$, etc.). This defines $r_1$ on 0-simplexes.

In general $r_1$ is defined by induction and the fact that all the choices made above were only within prescribed homotopy classes implies that a choice over the $(k-1)$-skeleton extends to the $k$-skeleton.

**Definition of $r_2 : (G/PL)_M \longrightarrow NM(M)$.** We now define a map $r_2$. $r_1$ and $-r_2$ are homotopy inverses ($-r_2$ means $r_2$ composed with inversion in $(G/PL)_M$, which, as an $H$-space, possesses a homotopy inverse.) It will be easy to verify that $r_1$ and $-r_2$ are in fact homotopy inverses and we leave this verification to the reader.

**0–simplexes.**

Let $(\zeta, g_1)$ be a 0-simplex of $(G/PL)_M$, meaning that $\zeta/M$ is a $PL$ bundle and $g_1 : \zeta/M \longrightarrow M \times \mathbb{R}^N$ is a fibre homotopy trivialization. Adding $\nu_M$ to both sides we have a fibre homotopy equivalence $g : \xi \longrightarrow \nu_M$, where $\xi = \nu_M \oplus \zeta$, and hence a homotopy equivalence $T(g) : T(\xi) \longrightarrow T(\nu_M)$. Thus the canonical reduction $t' : I^{n+N} \longrightarrow T(\nu_M)$ determines a reduction $t$ of $T(\xi)$. By Rourke and Sanderson [31] and Williamson [45] we may assume that $t$ is transverse regular to $M$ and hence $t^{-1}(M)$ is a $PL$ manifold $Q$ and $t|_Q : Q \longrightarrow M$ is covered by a map $t| : \nu_Q \longrightarrow \xi$ of bundles. This defines a 0-simplex of $NM(M)$.

In general the same argument applies to define $r_2$ on $k$-simplexes extending a given definition on $(k-1)$-simplexes. The only change needed is that one uses the relative transverse regularity theorem.

**Proof of Theorem 3.2.**

**Definition of $i$.**

**0-simplexes.** Let $h : (Q, \partial Q) \longrightarrow (M, \partial M)$ be a homotopy equivalence and let $\nu_Q$ denote the normal bundle of $Q$. Let $h'$ be a homotopy inverse of $h$ and let $\xi/M$ be $(h')^* \nu_Q$ then $h$ is covered by a bundle map $\widehat{h} : E(\nu_Q) \longrightarrow E(\xi)$. This defines a 0-simplex of $NM(M)$.

Again, since all choices were within canonical classes, this definition on 0-simplexes yields an inductive definition on $k$-simplexes. Notice that $i$ is an embedding. Thus to prove $i$ is a homotopy equivalence we only have to prove that $NM(M)$ deformation retracts on $HT(M)$. We prove the following assertion:

**Assertion.** Suppose $f : Q^{n,k} \longrightarrow M^n \times \Delta^k$ is a degree 1 map (covered by the usual bundle map, as always) and suppose $f|_{Q_\sigma}$ is a homotopy equivalence for each proper face $\sigma < \Delta^k$. Then $f$ is bordant rel $\bigcup_{\sigma \in \partial \Delta^k} Q_\sigma$ to a homotopy equivalence, (and the bordism is covered by the usual bundle map, extending the given map over $Q^{n,k}$).

The assertion implies that a typical relative homotopy element is zero and hence the result (for more detail see Rourke and Sanderson [33]).

**Proof of the assertion.** The bordism is constructed as the trace of a finite number of surgeries of $(Q_{\Delta^k}, (\partial Q)_{\Delta^k})$ (each surgery being covered by a map of bundles). The construction here is familiar (see Browder [1], Novikov [28], Wall [42]) so we shall not repeat it. Here is a lemma, proved in Wall [44, 1.1] which allows one to do the surgery.

**(3.9) Lemma.** *Every element $\alpha \in \ker(\pi_i(Q_{\Delta^k}) \longrightarrow \pi_i(M \times \Delta^k))$ gives rise to a well-defined regular homotopy class of immersions of $T = S^i \times D^{n+k-i}$ in $Q$. We can use an embedding of $T$ in $Q$ to perform surgery on $\alpha$ iff the embedding lies in this class.*

By 3.9 we can immediately perform surgery up to just below the middle dimension. To kill middle-dimensional classes use the method of surgery of relative classes in Wall [42, 44]. (For a direct proof that surgery obstructions are zero on a boundary see Rourke and Sullivan [36].) This completes 3.2.

**Proof of 3.4.** Observe that $c$ is an embedding and so we have to construct a deformation retract of $HT(M)$ on $c(HT(M))$. To construct this on 0-simplexes one has to prove that any homotopy triangulation is homotopic to one which is a $PL$ homeomorphism on the inverse image of a disc $D^n \subset M^n$. This is easy. In general we have to prove the following assertion, which follows from the splitting theorem of Browder [2].

**Assertion.** Suppose $f : Q^{n,k} \longrightarrow M^n \times \Delta^k$ is a $(\Delta^k, n)$-homotopy equivalence and $M^n$ is closed. Suppose $f|_{f^{-1}(D^n \times \partial\Delta^k)}$ is a $PL$ homeomorphism. Then $f$ is homotopic rel $\bigcup_{\sigma \in \partial\Delta^k} Q_\sigma$ to a map which is a $PL$ homeomorphism on $f^{-1}(D^n \times \Delta^k)$.

**Proof of 3.7.** $q : HT(M) \longrightarrow (G/PL)^M$ is defined as before. Suppose $h_1 : M_1 \longrightarrow M$ and $h_2 : M_2 \longrightarrow M$ are vertices in $HT(M)$ which map into the same component of $(G/PL)^M$. Then the proof of 3.1 yields a cobordism between $M_1$ and $M_2$ (covered by the usual map of bundles). By taking bounded connected sum with a suitable Kervaire manifold (cf. Browder and Hirsch [5]) we may assume that the surgery obstruction vanishes and hence this cobordism may be replaced by an $h$-cobordism. So $h_1$ and $h_2$ lie in the same component of $HT(M)$.

**Two remarks on $q_*$.**

(1) $HT(M)$ and $(G/PL)^{M_0}$ are both based sets and, examining the proofs of 3.1 and 3.2, we see that both $r$ and $i$ can be chosen to preserve base-points. So $q_*$ is base-point preserving and we can rephrase 3.6 and 3.7, as follows.

**(3.10) Corollary.** *Given a homotopy equivalence $h : M_1 \longrightarrow M$ there is defined, up to homotopy, a map $q_h : M_0 \longrightarrow G/PL$ with the property that $q_h \simeq *$ iff $h$ is homotopic to a PL isomorphism (in case $n$=5, interpret $M_0$ as $M$).*

(2) Suppose $h : M_1 \longrightarrow M$ is a (topological) homeomorphism then $h$ determines a topological isomorphism $\nu(h) : \nu_{M_1} \longrightarrow \nu_M$. This can be seen as follows. Regard $M$ as imbedded in a large-dimensional cube $I^{N+n}$. Use Gluck [10] to ambient isotope $h$ to a $PL$ embedding; reversing this isotopy and using the (stable) uniqueness theorem for topological normal bundles (Hirsch [15] and Milnor [26]) we obtain the required isomorphism of $\nu_{M_1}$ with $\nu_M$.

Now $\nu(h)$ determines a topological trivialization $t(h)$ of $[\nu_M] - [(h^{-1})^*(\nu_{M_1})]$ and hence a map $t_h : M \longrightarrow TOP/PL$ by 2.2.

**Proposition 3.11.** *Thc diagram*

$$\begin{array}{ccc} M & \xrightarrow{\quad t_h \quad} & TOP/PL \\ & {\scriptstyle q_h} \searrow \quad \swarrow {\scriptstyle i} & \\ & G/PL & \end{array}$$

*commutes up to homotopy.*

**Proof.** From the definition of $\nu(h)$ is clear that

$$\begin{array}{ccc} & (I^{n+N}, \partial I^{n+N}) & \\ {\scriptstyle u_1} \swarrow & & \searrow {\scriptstyle u} \\ (T(\nu_{M_1}), T(\nu_{\partial M_1})) & \xrightarrow{\quad T(\nu_h) \quad} & (T(\nu_M), T(\nu_{\partial M})) \end{array}$$

commutes up to homotopy, where $u_1$ and $u$ are the canonical reductions.

Comparing with diagram 3.8 we see that $\nu(h)$ coincides, up to fibre homotopy, with $g$ and hence the fibre homotopy trivialization of $[\nu_M] - [(h^{-1})^*(\nu_{M_1})]$ which determines $q_h$ (see below 3.8) coincides up to fibre homotopy with $t(h)$, as required.

## §4. Surgery obstructions

In this section we define the "canonical surgery obstruction"

$$\sigma \; : \; (G/PL)^M \longrightarrow \Omega^n(G/PL)$$

and the periodicity map

$$\mu\ :\ (G/PL) \longrightarrow \Omega^{4n}(G/PL)$$

mentioned in the introduction.

Throughout the section $M$ denotes a closed $PL$ $n$-manifold, $n \geq 6$, and $M_0 = M\backslash\{\text{disc}\}$.

**(4.1) Proposition.** *The restriction map* $p : (G/PL)^M \longrightarrow (G/PL)^{M_0}$ *is a fibration with fibre* $\Omega^n(G/PL)$.

**Proof.** To prove that $p$ has the homotopy lifting property one has to prove that a map of $M \times I^n \cup M_0 \times I^n \times I \longrightarrow G/PL$ extends to $M \times I^n \times I$ (see §2 for the notion of a map of a polyhedron in a $\Delta$-set) and this follows at once from the generalized extension condition proved in Rourke and Sanderson [33]. Thus the result will follow if we know that $p$ is onto, it is clear that the fibre will be $\Omega^n(G/PL)$.

Consider the diagram

$$\begin{array}{ccc} HT(M) & \xrightarrow{\ q'\ } & (G/PL)^M \\ \simeq \big\uparrow c & & p \big\downarrow \big\uparrow s \\ HT(M_0) & \xrightarrow[\simeq]{\ q\ } & (G/PL)^{M_0} \end{array}$$

$c$ and $q$ were defined in §3 and $q'$ is defined exactly as $q$. Then $q \simeq pq'c$ straight from the definitions. Let $s = q'cq^{-1}$ where $q^{-1}$ is a (homotopy) inverse to $q$, then $ps \simeq id$ and the result follows.

**(4.2) Theorem.** $(G/PL)^M$ *is homotopy equivalent to the product* $\Omega^n(G/PL) \times (G/PL)^{M_0}$.

**Proof.** Regard $\Omega^n(G/PL)$ as a subset of $(G/PL)^M$ by inclusion as the fibre. Define

$$d\ :\ \Omega^n(G/PL) \times (G/PL)^{M_0} \longrightarrow (G/PL)^M\ ;\ (x, y) \longmapsto \kappa m(\kappa^{-1}x, \kappa^{-1}sy)$$

where $\kappa^{-1}$ is a homotopy inverse of $\kappa$ and $m$, $\kappa$ are as defined in §2 (below 2.3).

We assert that the diagram

$$\begin{array}{ccccc} & & \Omega^n(G/PL) & & \\ & \swarrow \subseteq & & \subseteq \searrow & \\ \Omega^n(G/PL) \times (G/PL)^{M_0} & & \xrightarrow{\quad d \quad} & & (G/PL)^M \\ & \pi_2 \searrow & & \swarrow p & \\ & & (G/PL)^{M_0} & & \end{array}$$

commutes up to homotopy. The result follows by comparing the two exact homotopy sequences, using the five-lemma.

The top triangle commutes since $m(\ ,*) \simeq \mathrm{id}$. To prove that the bottom triangle commutes, let $x^k \in \Omega^n(G/PL)$ be a $k$-simplex. Then $\kappa^{-1}x$ is a pair $(\xi, t)$, where $\xi/M \times \Delta^k$ is a $PL$ bundle and $t$ is a fibre homotopy trivialization. Since $x$ lies in $\Omega^n(G/PL)$, $t$ is the identity on $\xi|_{M_0 \times \Delta^k}$. Now $m(\kappa^{-1}x, \kappa^{-1}sy)$ is the pair $(\xi \oplus \xi_1, t \oplus t_1)$ where $\xi_1/M \times \Delta^k$ is another bundle. Moreover

$$p_1 m(\kappa^{-1}x, \kappa^{-1}sy) \;=\; p_1 m(*^k, \kappa^{-1}sy) \ ,$$

where $p_1 : (G/PL)_M \longrightarrow (G/PL)_{M_0}$ is induced by restriction. Since it is clear that $p, p_1$ commute with $\kappa$, the result follows.

**Definition.** The composite

$$\sigma \,:\, (G/PL)^M \xrightarrow{\ d^{-1}\ } \Omega^n(G/PL) \times (G/PL)^{M_0} \xrightarrow{\ \pi_1\ } \Omega^n(G/PL)$$

is the **canonical surgery obstruction**, where $d^{-1}$ is some homotopy inverse to $d$.

We now recall the more usual surgery obstructions. The connection with $\sigma$ will be established in 4.4.

**The surgery obstruction of a class $\alpha \in [M, G/PL]$.**

According to 3.1 $\alpha$ can be interpreted as a bordism class of normal maps:

$$\begin{array}{ccc} E(\nu_{M_1}) & \xrightarrow{\ \widehat{f}\ } & E(\xi) \\ \downarrow & & \downarrow \\ M_1^n & \xrightarrow{\ f\ } & M^n \end{array}$$

One can then associate to $\alpha$ a surgery obstruction in the following groups:

$$s(\alpha) = \begin{cases} I(M) - I(M_1) \in 8\mathbb{Z} & n = 4k \\ 0 & \text{if } n \text{ is odd} \\ K(f) \in \mathbb{Z}_2 & \text{if } n = 4k+2\,. \end{cases}$$

Here $I(\ )$ denotes index and $K(\ )$ the Kervaire obstruction. Direct definitions of $K(f)$ are given by Browder [4] and Rourke and Sullivan [36]. The methods of Browder [1] and Novikov [28], translated into the $PL$ category, imply that $s(\alpha) = 0$ iff the bordism class of $(f, \widehat{f})$ contains a homotopy equivalence, $n \geq 5$.

**Computation of $\pi_n(G/PL)$.**

If $M^n$ is the sphere $S^n$, then $\xi_0 = \xi|_{S^n\setminus\{disc\}}$ is trivial and so $\nu_{M_1}|_{M_1\setminus\{disc\}}$ is trivial and in fact has, up to equivalence, a well-defined trivialization given by trivializing $\xi_0$. This recovers the theorem (see also Rourke and Sanderson [32] and Sullivan [41]):

**(4.3) Theorem.** *$\pi_n(G/PL) \cong P_n$, the group of almost framed cobordism classes of almost framed $PL$ $n$-manifolds.*

The surgery obstruction give maps $s : \pi_n(G/PL) \longrightarrow 8\mathbb{Z}, 0$, or $\mathbb{Z}_2$ which are injective for $n \geq 5$ by the Browder-Novikov theorem quoted above (using the Poincaré theorem). Moreover in this range $s$ is surjective, since all obstructions are realized by suitable Kervaire or Milnor manifolds (see Kervaire [18] and Milnor [25]). So we have

$$\pi_n(G/PL) = \begin{cases} 8\mathbb{Z} & \text{if } n = 4k \\ 0 & \text{if } n \text{ is odd} \\ \mathbb{Z}_2 & \text{if } n = 4k+2. \end{cases}$$

To compute $\pi_n(G/PL)$ for $n < 5$ it is necessary to use the braid of the triple $O \subset PL \subset G$ (see Levine [23], also Rourke and Sanderson [32]) and known homotopy groups. Then the above formulae hold for $n < 5$ as well. However there is a distinct singularity because the generator of $P_4$ has index 16 (Rohlin [29]) instead of 8 for $P_{4k}$, $k > 1$ (cf. Milnor [26]).

We now prove:

**(4.4) Theorem.** *The map $\sigma : (G/PL)^M \longrightarrow \Omega^n(G/PL)$ induces the surgery obstruction function*

$$\sigma_* = s : [M, G/PL] = \pi_0(G/PL)^M \longrightarrow \pi_0(\Omega^n(G/PL)) = \pi_n(G/PL) = P_n$$

*for $n \geq 5$.*

**Proof.** We first make two observations.

(1) The surgery obstruction is additive (under connected sums). The connected sum of normal maps $f : M_1 \longrightarrow M, g : Q_1 \longrightarrow Q$ is a normal map $f\#g : M_1\#Q_1 \longrightarrow M\#Q$ with surgery obstruction

$$s(f\#g) = s(f) + s(g) .$$

(2) The action of an element in $\pi_0(\Omega^n(G/PL))$ on an element in $[M, G/PL]$ given by multiplication in $(G/PL)^M$ corresponds to taking connected sum of the associated normal maps. (For the normal map corresponding to a vertex in $\Omega^n(G/PL)$ – after inclusion in $(G/PL)^M$ – is the identity outside a disc in $M$, and we can assume that the other situation is the identity in this disc.)

Now let $f : M \longrightarrow G/PL$ be a vertex of $(G/PL)^M$ and let $\alpha \in \pi_n(G/PL)$ be the class $-s(f)$. Let $\alpha_0 \in \Omega^n(G/PL)$ be a corresponding vertex. Then

$$\sigma_*[m_1(\alpha^0, f)] = [m_2(\sigma\alpha^0, \sigma f)] = \sigma_*[\alpha^0] + \sigma_*[f] ,$$

with $m_1$ and $m_2$ the multiplications in $(G/PL)^M$ and $\Omega^n(G/PL)$ respectively. But

$$[m_1(\alpha^0, f)] = [*]$$

by choice of $\alpha$ and observations (1) and (2), so that $\sigma_*[f] = -\sigma_*[\alpha_0]$. The composite

$$\Omega^n(G/PL) \subset (G/PL)^M \xrightarrow{\sigma} \Omega^n(G/PL)$$

is the identity (by definition); hence $\sigma_*[\alpha^0] = \alpha$ and the result follows.

**Remarks.** (1) 4.4 fails for $n = 4$ (we cannot even define $\sigma$ in this case). Indeed if one considers the composition

$$\pi_4(G/PL) \xrightarrow{i_*} [M^4, G/PL] \xrightarrow{s} \mathbb{Z}$$

($s$ denotes the surgery obstruction) then $si_*(\pi_4(G/PL)) = 16\mathbb{Z}$, as remarked above, while $s$ maps **onto** $8\mathbb{Z}$ for suitable choice of $M$. This follows from the fact that $24\mu/\mathbb{CP}^2$ is fibre homotopy trivial, where $\mu$ is the normal bundle of $\mathbb{CP}^2$ in $\mathbb{CP}^3$. Then the surgery obstruction of the corresponding map of $\mathbb{CP}^2$ into $G/PL$ is 24(Hirzebruch index of $\mu$) $= 24$.
(2) It has been obvious for some time that $\pi_n(G/PL) \longrightarrow \Omega_n^{PL}(G/PL)$, where $\Omega_n^{PL}(\ )$ denotes $PL$ bordism, is a monomorphism (see e.g. observation (2) above). In fact it follows at once from 4.4 (and the fact that surgery obstructions are bordism invariants) that $\pi_n(G/PL)$ splits as a direct summand of $\Omega_n^{PL}(G/PL)$ for $n \geq 5$ (see also Sullivan [41]). This remark also fails for $n = 4$ for the same reasons as remark (1).

**The periodicity map.**

We now define the periodicity map mentioned in the introduction.

**Definition.** $\mu$ is the composite

$$\mu : G/PL \xrightarrow{\alpha} (G/PL)^{\mathbb{CP}^{2n}} \xrightarrow{\sigma} \Omega^{4n}(G/PL)$$

where $\alpha$ is a semi-simplicial approximation to the map (of spaces) defined in the introduction (see Rourke and Sanderson [33]).

**(4.5) Theorem.** *The map* $\mu_* : \pi_i(G/PL) \longrightarrow \pi_i(\Omega^{4n}(G/PL)) = \pi_{i+4n}(G/PL)$ *is an isomorphism for* $i \neq 4$ *and is the inclusion* $16\mathbb{Z} \longrightarrow 8\mathbb{Z}$ *for* $i = 4$.

**Proof.** By Theorem 4.4 we have to consider the composition

$$[S^i, G/PL] \xrightarrow{\alpha_*} [S^i \times \mathbb{CP}^{2n}, G/PL] \xrightarrow{s} \pi_{i+4n}(G/PL)$$

As mentioned earlier, an element $\beta \in [S^i, G/PL]$ is represented by a normal map

$$\begin{array}{ccc} E(\nu_M) & \xrightarrow{\hat{f}} & E(\xi) \\ \downarrow & & \downarrow \\ M^i & \xrightarrow{f} & S^i \end{array}$$

and then $\alpha_*(\beta)$ is represented by the normal map

$$\begin{array}{ccc} E(\nu_M \times \nu_{\mathbb{CP}^{2n}}) & \xrightarrow{\hat{f} \times 1} & E(\xi \times \nu_{\mathbb{CP}^{2n}}) \\ \downarrow & & \downarrow \\ M^i \times \mathbb{CP}^{2n} & \xrightarrow{f \times 1} & S^i \times \mathbb{CP}^{2n} \end{array}$$

(this is easily checked from the proof of 3.1). So it is necessary to know how surgery obstructions behave under cartesian product. A complete answer is provided by Rourke and Sullivan [36]. Using the fact that $I(\mathbb{CP}^{2n}) = 1$ the required result follows.

## §5. The 'canonical Novikov homotopy'

We now complete the proof of the Main Theorem (1.1) by proving:

**(5.1) Theorem.** *The composite*

$$(TOP/PL)^M \xrightarrow{i_*} (G/PL)^M \xrightarrow{\sigma} \Omega^n(G/PL)$$

*is null-homotopic, assuming $M$ is closed of dimension $n \geq 5$ .*

**Proof.** We construct a map $\lambda : (TOP/PL)^M \longrightarrow HT(M)$ such that

$$\begin{array}{ccc} (TOP/PL)^M & \xrightarrow{\quad i_* \quad} & (G/PL)^M \\ & \lambda \searrow \quad \nearrow q' & \\ & HT(M) & \end{array} \tag{5.2}$$

is homotopy commutative. But by the definition of $\sigma$ we have

$$\begin{array}{ccc} HT(M) & \xrightarrow{\quad q' \quad} & (G/PL)^M \\ {\scriptstyle (\sigma^{-1}q')\times *}\downarrow & & \downarrow{\scriptstyle \sigma} \\ (G/PL)^{M_0} \times \Omega^n(G/PL) & \xrightarrow{\quad \pi_2 \quad} & \Omega^n(G/PL) \end{array}$$

homotopy commutative, so that $\sigma q' \simeq *$ and the theorem follows.

**Construction of $\lambda$.** In fact we shall construct a map $\lambda_1 : (TOP/PL)_M \longrightarrow HT(M)$ so that

$$\begin{array}{ccc} (TOP/PL)_M & \xrightarrow{\quad j \quad} & (G/PL)_M \\ & \lambda_1 \searrow \quad \nearrow q_1 & \\ & HT(M) & \end{array} \tag{5.3}$$

commutes up to homotopy, where $q_1 = r_1 \circ i$ (see §3). The result then follows by 2.3.

The main tool in the construction of $\lambda_1$ is a refined version of the Novikov-Siebenmann splitting theorem. In what follows all maps of bounded manifolds are assumed to carry boundary to boundary.

**Definition.** Suppose $h : W \longrightarrow M \times \mathbb{R}^k$ is a topological homeomorphism, $W$ and $M$ being $PL$ manifolds. We say $h$ **splits** if there is a $PL$ isomorphism $g : M_1 \times \mathbb{R}^k \longrightarrow W$ such that the composite $hg : M_1 \times \mathbb{R}^k \longrightarrow M \times \mathbb{R}^k$ is properly homotopic to $f \times \mathrm{id}_{\mathbb{R}^k}$, where $f$ is the composite

$$M_1 \subset M_1 \times \mathbb{R}^k \xrightarrow{hg} M \times \mathbb{R}^k \xrightarrow{\pi_1} M .$$

A splitting of $h$ is a triple $(M_1, g, H)$ where $M_1$, $g$ are as above and $H$ is a proper homotopy between $hg$ and $f \times \mathrm{id}_{\mathbb{R}^k}$. Two splittings $(M_1, g, H)$ and $(M_1', g', H')$

are **equivalent** if there is a $PL$ isomorphism $e : M_1 \longrightarrow M_1'$ such that diagrams (a) and (b) commute up to isotopy and proper homotopy respectively:

$$\begin{array}{ccc} M_1 \times \mathbb{R}^k & \xrightarrow{\quad g \quad} & W \\ {\scriptstyle e \times \{\mathrm{id}\}} \searrow & & \nearrow {\scriptstyle g'} \\ & M_1' \times \mathbb{R}^k & \end{array}$$

(a)

$$\begin{array}{ccc} M_1 \times \mathbb{R}^k \times I & \xrightarrow{\quad H \quad} & M \times \mathbb{R}^k \\ {\scriptstyle e \times \{\mathrm{id}\}} \searrow & & \nearrow {\scriptstyle H'} \\ & M_1' \times \mathbb{R}^k \times I & \end{array}$$

(b)

**(5.4) Theorem.** *Suppose given $h$ as above and $M$ is closed of dimension $\geq 5$ or bounded of dimension $\geq 6$ with $\pi_1(M) = \pi_1(\partial M) = 0$. Then there is a well-defined equivalence class of splittings of $h$ which we call 'Novikov splittings'. If $M$ is bounded then the restriction of a Novikov splitting of $h$ to $\partial M$ is a Novikov splitting of $h|_{\partial M}$.*

**Remark.** The second half of 5.4 implies that any Novikov splitting of $h|_{\partial W}$ extends to a Novikov splitting of $h$.

Theorem 5.4 is proved by constructing a **tower** of interpolating manifolds for

$$h| \; : \; h^{-1}(M \times T^{k-1} \times \mathbb{R}) \longrightarrow M \times T^{k-1} \times \mathbb{R}$$

(cf. Novikov [27]), and applying inductively Siebenmann's 1-dimensional splitting theorem [37] (translated into the $PL$ category using the techniques of Rourke and Sanderson [30, 31]). Full details are to be found in Lashof and Rothenberg [22] or Chapter I of these notes.

**Definition of $\lambda_1$ on 0-simplexes.**

Suppose $\sigma^0 \in (TOP/PL)_M$ is a 0-simplex. Then $\sigma^0$ is a pair $(\xi, h)$ where $\xi^k/M$ is a $PL$ bundle and $h : E(\xi^k) \longrightarrow M \times \mathbb{R}^k$ is a topological trivialization. Let

$g : M_1 \times \mathbb{R}^k \longrightarrow E(\xi)$ be a Novikov splitting for $h$. Then the composite $f : M_1 \longrightarrow M$ (as in the definition above) is a homotopy equivalence and hence a 0-simplex $\lambda_1(\sigma^0) \in HT(M)$.

In general $\lambda_1$ is defined by induction on dimension. The general definition is similar to that for 0-simplexes except that one uses the relative version of 5.4.

**Commutativity of 5.3.**

In §3 $q_1 = r_1 \circ i$ was only defined up to homotopy. We shall prove that the definition of $q_1$ could have been chosen so that (5.3) commutes precisely. We prove this for 0-simplexes. The general proof is similar.

Let $\sigma^0 \in (TOP/PL)_M$ as above. Then we have:

$$\begin{array}{ccc} M_1 \times \mathbb{R}^k & \xrightarrow[\cong]{\quad g \quad} & E(\xi) \\ & {\scriptstyle f \times \{\mathrm{id}\}} \searrow \quad \swarrow {\scriptstyle h} & \\ & M \times \mathbb{R}^k & \end{array} \tag{5.5}$$

commuting up to proper homotopy. Add the (stable) bundle $\nu_{M_1}$ to all the terms in (5.5) and observe that $(g^{-1})^*(\nu_{M_1}) \oplus \xi$ is the stable normal bundle of $M$ since its total space (which is the same as the total space of $\nu_{M_1}$) is embedded in a sphere. We obtain

$$\begin{array}{ccc} E(\nu_{M_1}) & \xrightarrow[\cong]{\quad g_1 \quad} & E(\nu_M) \\ & {\scriptstyle \widehat{f}} \searrow \quad \swarrow {\scriptstyle \widehat{h}} & \\ & E((f^{-1})^*\nu_{M_1}) & \end{array}$$

The pair $(f|_{M_1}, \widehat{f})$ is a normal map of $M_1$ to $M$ which we can take to be $i(\sigma^0)$ (see definition of $i$ in §3).

Now it is clear that $g_1$ commutes (up to homotopy) with the canonical reductions of $T(\nu_{M_1})$ and $T(\nu_M)$. Consequently $\widehat{h}$ can be taken to be the fibre homotopy equivalence $\nu_M \longrightarrow (f^{-1})^*(\nu_{M_1})$ determined by $\widehat{f}$. (Cf. diagram (3.8) et seq.) $\widehat{h}$ determines a fibre homotopy trivialization of $[\xi] = [\nu_M] - [(f^{-1})^*(\nu_{M_1})]$ which we may take to be $h$ itself. So we may take $q_1(\sigma^0) = (\xi, h)$, as required.

## §6. Weaker hypotheses

Here we sketch a proof that the condition (3) in the Main Theorem (1.1) can be weakened to

(3′) $H^4(M; \mathbb{Z})$ *has no elements of order 2.*

Full details of the proof are contained in chapter III of these notes.

The idea of the proof is this. We proved that the Sullivan obstruction $q_h : M_0 \longrightarrow G/PL$ corresponding to the homeomorphism $h : Q \longrightarrow M$ factored via the fibre $K(\mathbb{Z}_2, 3)$ of $\mu : G/PL \longrightarrow \Omega^{4n}(G/PL)$. This factoring is not unique, it can be altered by multiplication (in $K(\mathbb{Z}_2, 3)$) with any map of $M_0$ into the fibre of $K(\mathbb{Z}_2, 3) \longrightarrow G/PL$ which is $\Omega^m(G/PL)$, $m = 4n + 1$. We shall show that a suitable map of $M_0$ can be chosen so that the obstruction is changed by the mod 2 reduction of any class in $H^3(M, \mathbb{Z})$. Then consider the exact coefficient sequence:

$$H^3(M;\mathbb{Z}) \xrightarrow{\text{mod } 2} H^3(M;\mathbb{Z}_2) \xrightarrow{\beta} H^4(M;\mathbb{Z}) \xrightarrow{\times 2} H^4(M;\mathbb{Z})\,.$$

If condition (3′) holds, $\beta$ is zero, and the entire obstruction can be killed.

To prove that a suitable map of $M_0$ into $\Omega^m(G/PL)$ can be found, it is necessary to examine the structure of $G/PL$ for the prime 2.

**Definition**. Suppose $X$ is an $H$-space and $R$ is a subring of the rationals. $X \otimes R$ is a $CW$ complex which classifies the generalized cohomology theory $[\ \ , X] \otimes R$ (see Brown [6]).

The ring $\mathbb{Z}_{(2)}$ of integers localized at 2 is the subring of the rationals generated by $\frac{1}{p_i}$ with $p_i$ the odd primes. We write $X_{(2)} = X \otimes \mathbb{Z}_{(2)}$.

**(6.1) Theorem**. *The $k$-invariants of $(G/PL)_{(2)}$ are all trivial in dimension* $\geq 5$.

Assume 6.1 for the moment. To prove our main assertion we deduce:

**(6.2) Corollary**. $\Omega^m(G/PL)_{(2)}$ $(m = 4n + 1, n > 0)$ *is homotopy equivalent to the cartesian product of* $K(\mathbb{Z}_2, 4i + 1)$ *and* $K(\mathbb{Z}_{(2)}, 4i - 1)$, $i = 1, 2, \ldots$.

Next we assert that the composite

$$K(\mathbb{Z}_{(2)}, 3) \subset \Omega^m(G/PL)_{(2)} \longrightarrow K(\mathbb{Z}_2, 3)$$

is "reduction mod 2". This follows from the observation that, from the homotopy properties of $\mu$, the map $\Omega^m(G/PL) \longrightarrow K(\mathbb{Z}_2, 3)$ is essential. Now let $\alpha \in H^3(M;\mathbb{Z})$ be any class and let $\alpha_1 \in H^3(M;\mathbb{Z}_{(2)})$ be the corresponding class. Let $\alpha_2 \in H^3(M;\mathbb{Z}_2)$ be the reduction mod 2 of $\alpha_1$ (and $\alpha$). $\alpha_1$ is realized by a map $f : M_0 \longrightarrow K(\mathbb{Z}_{(2)}, 3) \subset \Omega^m(G/PL)_{(2)}$ and some odd multiple $rf$ lifts to $\Omega^m(G/PL)$. But, on composition into $K(\mathbb{Z}_2, 3)$, $rf$ also represents $\alpha_2$, and so we can indeed alter the original obstruction by the mod 2 reduction of $\alpha$, as asserted.

**Proof of 6.1**. The main step is the construction of cohomology classes in $H^{4*}(G/PL; \mathbb{Z}_{(2)})$ and $H^{4*+2}(G/PL; \mathbb{Z}_2)$ which determine the surgery obstructions:

**(6.3) Theorem.** *There are classes*

$$\mathcal{K} = k_2 + k_6 + \ldots \in H^{4*+2}(G/PL;\mathbb{Z}_2)$$

*and*

$$\mathcal{L} = \ell_4 + \ell_8 + \ldots \in H^{4*}(G/PL;\mathbb{Z}_{(2)})$$

*with the following property. If* $f : M^n \longrightarrow G/PL$ *is a map then*

$$s(f) = \begin{cases} \langle W(M) \cup f^*\mathcal{K}, [M]\rangle \in \mathbb{Z}_2 & \text{if } n = 4k+2 \\ 8\langle L(M) \cup f^*\mathcal{L}, [M]\rangle \in 8\mathbb{Z} & \text{if } n = 4k \end{cases}$$

*where* $W(M)$ *is the total Stiefel-Whitney class and* $L(M)$ *is the Hirzebruch L-genus.*

**Remark** $L(\ )$ is a rational class (obtained from the equivalence $BO \otimes \mathbb{Q} \simeq BPL \otimes \mathbb{Q}$, which follows from the finiteness of the exotic sphere groups $\Theta_i = \pi_i(PL/O)$ (Kervaire and Milnor [19]). The second formula must the therefore be interpreted in rational cohomology.

**(6.4) Corollary** *If* $f : S^m \longrightarrow G/PL$ *represents the generator of* $\pi_{4n}$ *(resp.* $\pi_{4n+2}$*),* $n > 1$*, then*

$$\langle f^*(\ell_{4n}), [S^{4n}]\rangle = 1 \ \ (resp.\ \langle f^*(k_{4n+2}), [S^{4n+2}]\rangle = 1\ ) .$$

It follows from 6.4 that the Hurewicz map for $(G/PL)_{(2)}$ is indivisible in dimensions $4n$, $n > 1$, and that the mod 2 Hurewicz map is non-trivial in dimensions $4n+2$. From these facts, 6.1 follows by an exact sequence argument.

We now prove 6.3.

**Definition of $\mathcal{K}$.** Assuming that

$$\mathcal{K}^{r-1} = k_2 + k_6 + \ldots + k_{4r-2}$$

has already been defined we define $k_{4r+2}$. By Thom (see Conner and Floyd [8]) we have that

$$\Omega_{4r+2}(G/PL;\mathbb{Z}_2) \longrightarrow H_{4r+2}(G/PL;\mathbb{Z}_2)$$

is onto (where $\Omega_*(\ )$ denotes oriented bordism), with kernel generated by decomposables. Let $x \in H_{4r+2}(G/PL;\mathbb{Z}_2)$ and $f : M \longrightarrow G/PL$ represent $x$. Define

$$k(x) = K(f) - \langle f^*\mathcal{K}^{r-1} \cup W(M), [M]\rangle \in \mathbb{Z}_2 .$$

Then from the multiplicative formulae for the Kervaire obstruction (Rourke and Sullivan [36]) and the multiplicative property of $W(\ )$, it is easy to check that $k(\ )$ vanishes on decomposables and therefore defines a cohomology class $k_{4r+2}$ with the required properties.

**Definition of $\mathcal{L}$.** The definition is very similar to $\mathcal{K}$. One uses instead the fact (also due to Thom) that

$$\Omega_{4r}(G/PL) \longrightarrow H_{4r}(G/PL; \mathbb{Z}_{(2)})$$

is onto, and the multiplicative properties of $L(\ )$ and the index obstruction.

This completes the proof of 6.3.

**Remark.** We have described as little of the homotopy type of $G/PL$ as we needed. Sullivan has in fact completely determined the homotopy type of $G/PL$. We summarize these results:

**At the prime 2.** $(G/PL)_{(2)}$ has one non-zero $k$-invariant (in dimension 4) which is $\delta Sq^2$ (this follows from 6.3 and the remarks below 4.4).

**At odd primes.** $(G/PL)_{(odd)}$ has the same homotopy type as $(BO)_{(odd)}$ (the proof of this is considerably deeper). [$X_{(odd)}$ means $X \otimes \mathbb{Z}[\frac{1}{2}]$. ]

## §7. Refinements of the Main Theorem

We consider three refinements:

(a) Relaxing the $\pi_1$-condition (2) in Theorem 1.1. No really satisfactory results are available here since one immediately meets the problem of topological invariance of Whitehead torsion.* However, if one is willing to bypass this problem and assume that $h$ is a simple homotopy equivalence, then one can relax condition (2) considerably in the bounded case.

(b) and (c) Relaxing the condition that $h$ is a homeomorphism. The two conditions we replace this by are:

(b) There is a topological $h$-cobordism between $M$ and $Q$.

(c) $h$ is a **cell-like** map (cf. Lacher [21]).

With both these replacements, the Main Theorem (1.1) holds good.

We first consider condition (a), assuming that $M^n$ is connected with non-empty connected boundary, $n \geq 6$, and $\pi_1(\partial M) \longrightarrow \pi_1(M)$ (induced by inclusion) is an isomorphism.

Let $SHT(M)$ denote the $\Delta$-set of simple homotopy triangulations of $M$, i.e. a typical $k$-simplex is a simple homotopy equivalence of pairs

$$(Q_{n,k}, \partial Q_{n,k}) \longrightarrow (M \times \Delta^k, \partial M \times \Delta^k) .$$

**(7.1) Theorem.** *$i : SHT(M) \longrightarrow NM(M)$ is a homotopy equivalence.*

**Proof.** The proof is the same as for 3.2, except that one deals with non-simply-

* Solved by Chapman in 1974.

connected surgery in the bounded case with the same fundamental group in the interior and on the boundary, so that the $\pi$-$\pi$ theorem of Wall [44] applies.

Combining 7.1 with 3.1 we have:

**(7.2) Corollary.** *$q : SHT(M) \longrightarrow (G/PL)^M$ is a homotopy equivalence.*

Now define $Sht(M)$ to be the set of $PL$ equivalence classes of simple homotopy triangulations of $M$. Then from the $s$-cobordism theorem we have

**(7.3) Corollary.** *$q_* : Sht(M) \longrightarrow [M, G/PL]$ is a bijection.*

Using 7.3 we now have precisely the same analysis as in the simply connected case and can deduce

**(7.4) Theorem.** *Suppose $h : Q \longrightarrow M$ is a homeomorphism* and *a simple homotopy equivalence. Suppose that $H_3(M; \mathbb{Z})$ has no 2-torsion. Then $h$ is homotopic to a $PL$ isomorphism.*

**(7.5) Corollary.** *If $h : Q \longrightarrow M^n$ is a topological homeomorphism, $n \geq 6$, and $M \searrow K^2$ then $h$ is homotopic to a $PL$ isomorphism.*

**Proof.** The dimension condition ensures that $\pi_1(\partial M) \longrightarrow \pi_1(M)$ is an isomorphism and Connell and Hollingsworth [7] show that $h$ must be a simple homotopy equivalence.

We now move on to condition (b).

**(7.6) Theorem.** *Suppose $M$ satisfies the conditions of the main theorem 1.1 with (2) replaced by the existence of a (topological) $h$-cobordism $W$ between $M$ and $Q$. Then the homotopy equivalence $h : Q \longrightarrow M$ determined by $W$ is homotopic to a $PL$ isomorphism.*

**Proof.** We only need to show that $q_h$ factors via $TOP/PL$. By Gluck [10] we may assume that $W$ is embedded properly in $S^N \times I$ with $PL$ embeddings $M \subset S^N \times \{0\}$, $Q \subset S^N \times \{1\}$ . By (stable) existence and uniqueness of normal bundles, we may assume that $W \subset S^N \times I$ has a normal bundle $\xi$ which restricts to $PL$ normal bundles $\nu_M$ and $\nu_Q$ on $M \subset S^N \times \{0\}, Q \subset S^N \times \{1\}$.

Since $W$ deformation retracts on $M$ and $Q$, $\xi$ is determined by each of $\nu_M$ and $\nu_Q$, therefore $(f^{-1})^*\nu_Q$ is topologically equivalent to $\nu_M$. But this equivalence clearly commutes with the standard reductions of Thom spaces and hence (cf. §3) coincides, up to fibre homotopy, with the fibre homotopy equivalence which determines $q_h$. Thus $q_h$ factors via $TOP/PL$, as required.

We now move on to condition (c).

**Definition**. *A map $f : Q \longrightarrow M$ of manifolds is* **cell-like** *(CL) if:*

(1) *$f$ is* **proper**, *i.e. $f^{-1}(\partial M) = \partial Q$ and $f^{-1}$ (compact)= compact; and one of the following holds:*

$(2)_1$ *$f^{-1}(x)$ has the Čech homotopy type of a point, for each $x \in M$*

$(2)_2$ *$f| : f^{-1}(U) \longrightarrow U$ is a proper homotopy equivalence, for each open set $U \subset M$ or $\partial M$.*

Lacher [21] proves equivalence of $(2)_1$ and $(2)_2$.

**(7.7) Theorem**. *Suppose $M$ satisfies the conditions of the main theorem and $f : Q \longrightarrow M$ is a cell-like map. Then $f$ is homotopic to a PL isomorphism.*

To prove 7.7 we shall define a $\Delta$-monoid $CL$ (analogous to $TOP$) and check that the same analysis holds.

**Definition of $CL_q$**. A typical $k$-simplex is a $CL$ fibre map $f : \Delta^k \times \mathbb{R}^q \longrightarrow \Delta^k \times \mathbb{R}^q$, i.e. $f$ commutes with projection on $\Delta^k$ and $f| : \{x\} \times \mathbb{R}^q \longrightarrow \{x\} \times \mathbb{R}^q$ is cell-like for each $x \in \Delta^k$.

The inclusion $CL_q \subset CL_{q+1}$ is defined by identifying $f$ with $f \times \mathrm{id}$ and the stable limit is $CL$.

Now redefine $G_q$ to consist of proper homotopy equivalences $\mathbb{R}^q \longrightarrow \mathbb{R}^q$ (clearly homotopy equivalent to our original definition) then we have $CL_q \subset G_q$ and $CL_q/PL_q \subset G_q/PL_q$.

Theorem 7.7 follows in the same way as the main theorem from the following three propositions:

**(7.8) Proposition**. *$CL/PL$ classifies stable PL bundles with a $CL$-trivialization (a $CL$-trivialization of $\xi/K$ is a fibre map $E(\xi) \longrightarrow K \times \mathbb{R}^q$ which is cell-like on fibres).*

**(7.9) Proposition**. *$q_f$ factors via $CL/PL$.*

**(7.10) Proposition**. *There is a map $\lambda_1 : (CL/PL)_M \longrightarrow HT(M)$ (M closed,* *simply connected and* $n \geq 6$*) which makes*

$$\begin{array}{ccc} (CL/PL)_M & \xrightarrow{\quad j \quad} & (G/PL)_M \\ & \searrow^{\lambda_1} \quad {}^{q_1}\nearrow & \\ & HT(M) & \end{array}$$

*homotopy commutative.*

Proposition 7.8 is best proved directly (an easy argument) rather than as the fibre of $BPL \longrightarrow BCL$, since it is not clear what $BCL$ classifies. We leave this to the reader. For 7.10 notice that the only fact used in defining $\lambda_1$ (cf. proof of 5.4) is that $h : E(\xi) \longrightarrow M \times \mathbb{R}^k$ is a proper homotopy equivalence on $h^{-1}(M \times T^{k-1} \times \mathbb{R})$, which is certainly implied if $h$ is cell-like. It remains to prove 7.9. For this we associate to $f : Q \longrightarrow M$ a $CL$ fibre map $\widehat{f} : \tau_Q \longrightarrow \tau_M$. Since the definition of $\widehat{f}$ is natural (induced by $f \times f : M \times M \longrightarrow Q \times Q$) it is easily proved that the $CL$ trivialization of $\tau_M \oplus (f^{-1})^* \nu_Q$ which $\widehat{f}$ determines, coincides up to fibre homotopy, with the fibre homotopy trivialization determined by $f$ as a homotopy equivalence (cf. §3 – one only needs to prove (3.8) commutative.).

Construct $\widehat{f}$ as follows : let

$$M \xrightarrow{\ \Delta\ } M \times M \xrightarrow{\ \pi_1\ } M$$

be the tangent microbundle of $M$. By Kuiper and Lashof [20], $\pi_1$ contains a $PL$ fibre bundle $\tau_M$ with zero section $\Delta_M$. Let

$$E_Q \ = \ (f \times f)^{-1} E(\tau_M) \subset Q \times Q \ .$$

We assert that $E_Q$ is the total space of a $PL$ fibre bundle, fibre $\mathbb{R}^n$, projection $\pi_1$ and zero-section $\Delta_Q$. To see this observe that

$$V \ = \ E_Q \cap \pi_1^{-1}(x) \cong f^{-1}(U)$$

where $U = \pi_2(\pi_1^{-1}(f_x) \cap E(\tau_M)) \subset M$, which is $\cong \mathbb{R}^n$. Now $f| : V \longrightarrow U$ is a proper homotopy equivalence, since $f$ is cell-like, so $V \cong \mathbb{R}^n$ by Stallings [39]. Hence $E_Q$ is an $\mathbb{R}^n$-bundle; $\Delta_Q$ is a section since $E_Q$ is a neighborhood of $\Delta_Q$ in $Q \times Q$ and we can take it as zero section. By the uniqueness part of [20], $E_Q \cong \tau_Q$ and $(f \times f)|_{E_Q}$ is the required $CL$ bundle map $\widehat{f}$.

In the bounded case, one first extends $f$ to a cell-like map of open manifolds (by adding an open collar to $M$ and $Q$ and extending $f$ productwise), then the above procedure works on restricting again to $Q$.

## §8. Block bundles and homotopy equivalences

We refer the reader to Rourke and Sanderson [30] for notions of block bundle etc. A **block homotopy equivalence** $h : E(\eta) \longrightarrow E(\xi)$, where $\xi^q, \eta^q/K$ are $q$-block bundles, is a map satisfying :

(1) $h|_K = \text{id}$ ,
(2) $h$ preserves blocks ,
(3) $h(E(\partial\eta)) \subset E(\partial\xi)$ ($\partial\xi, \partial\eta$ denote associated sphere block bundles)
(4) $h| : E(\partial\eta|_\sigma) \longrightarrow E(\partial\xi|_\sigma)$ has degree $\pm$ 1 each $\sigma \in K$.

Condition (3) makes sense since both sides have the homotopy type of $S^{q-1}$.

The proof of the following is left to the reader (cf. Dold [9]):

**(8.1) Proposition**. *There is an "inverse" block homotopy equivalence $g : E(\xi) \longrightarrow E(\eta)$ such that hg and gh are homotopic to the identity via block homotopy equivalences.*

**The associated $\widetilde{G}_q/\widetilde{PL}_q$-bundle.**

Define $\Delta$-sets $\widetilde{G}_q, \widetilde{PL}_q$ to consist of b.h.e.'s $\Delta^k \times I^q \longrightarrow \Delta^k \times I^q$ and block bundle isomorphisms $\Delta^k \times I^q \longrightarrow \Delta^k \times I^q$. These sets have the same homotopy type as the sets defined in §2 (cf. Rourke and Sanderson [32]). We associate to $\xi^q$ a $\Delta$-fibration with base $K$ and fibre $\widetilde{G}_q$ by taking as typical $k$-simplex a b.h.e. $f : \Delta^k \times I^q \longrightarrow E(\xi|_\sigma), \sigma^k \in K$, and a $\Delta$-fibration fibre $\widetilde{G}_q/\widetilde{PL}_q$ by factoring by $\widetilde{PL}_q$ on the left. I.e. $f_1 \sim f_2$ if there is $g \in \widetilde{PL}_q^{(k)}$ such that $f_1 = f_2 \circ g$.

Now say b.h.e.'s $h_1 : E(\eta_1) \longrightarrow E(\xi)$, $h_2 : E(\eta_2) \longrightarrow E(\xi)$ are **isomorphic** (resp. **homotopic**) if there is a block bundle isomorphism $g : \eta_1 \longrightarrow \eta_2$ such that $h_2 g = h_1$ (resp. $h_2 g$ is homotopic to $h_1$ via b.h.e.'s). The following is easily proved (cf. Rourke and Sanderson [30]) :

**(8.2) Proposition**. *Isomorphism classes (resp. homotopy classes) of b.h.e.'s $E(\eta) \longrightarrow E(\xi)$ correspond bijectively to cross-sections (resp. homotopy classes of cross-sections) of the associated $\widetilde{G}_q/\widetilde{PL}_q$-bundle to $\xi$.*

Now write $Ht(\xi)$, "homotopy triangulations of $\xi$", for the set of homotopy classes of b.h.e.'s $\eta \longrightarrow \xi$.

**(8.3) Corollary**. *If $q \geq 3$, $Ht(\xi) \cong [K, G/PL]$.*

**Proof**. This follows from 8.2 and 2.1(d), from the fact that $G/PL$ is an $H$-space, and from the existence of one cross-section (determined by id : $\xi \longrightarrow \xi$).

More generally define a $\Delta$-set $HT(\xi)$ with $\pi_0(HT(\xi)) = Ht(\xi)$ by taking as typical $k$-simplex an isomorphism class of b.h.e.'s $\eta \longrightarrow \xi \times \Delta^k$ (see Rourke and

Sanderson [30] for the cartesian product of block bundles), then one has similarly:

**(8.4) Corollary.** *If* $q \geq 3$, $HT(\xi) \simeq (G/PL)^K$.

Now suppose $|K| = M^n$ then $E(\xi)$ is a manifold and a block homotopy equivalence $\eta \longrightarrow \xi$ gives a simple homotopy equivalence $(E(\eta), E(\partial\eta)) \longrightarrow (E(\xi), E(\partial\xi))$, so we have a $\Delta$-map $j : HT(\xi) \longrightarrow SHT(E(\xi))$.

**(8.5) Theorem.** *If* $q \geq 3$, $n + q \geq 6$ *then* $j$ *is homotopy equivalence.*

**Proof.** By 8.4 and 7.2, both sides have the homotopy type of $(G/PL)^M$, so one only needs to check that the diagram

$$\begin{array}{ccc} HT(\xi) & \xrightarrow{\quad j \quad} & SHT(E(\xi)) \\ & {}_{8.4}\searrow \quad \swarrow_{q} & \\ & (G/PL)^M & \end{array}$$

commutes up to homotopy. Now 8.4 was defined by comparing $\xi$ and $\eta$ as (stable) block bundles and $q$ was defined (cf. §3) by comparing $\tau(E(\xi))$ and $\tau(E(\eta))$ as stable bundles. But $\tau(E(\xi)) \sim \xi \oplus \tau_M$ and $\tau(E(\eta)) \sim \eta \oplus \tau_M$ (see Rourke and Sanderson [31]) and it follows that the diagram commutes up to inversion in $G/PL$.

**Relative Sullivan theory.**

Suppose $Q \subset M$ is a codimension 0 submanifold and consider homotopy triangulations $h : M_1 \longrightarrow M$ which are $PL$ isomorphisms on $Q_1 = f^{-1}(Q) \subset M_1$. Denote the resulting $\Delta$-set $HT(M/Q)$, cf. §2.

The following is proved exactly as 3.3 and 3.5:

**(8.6) Theorem.** *There is a homotopy equivalence*

$$HT(M/Q) \;\simeq\; (G/PL)^{M_0/Q}$$

*if* $n \geq 6, \pi_1(M \backslash Q) = \pi_1(\partial M \backslash \partial Q) = 0$ *and* $M_0 = M$ *if* $\partial M \backslash \partial Q$ *is non-empty, and* $M_0 = M \backslash \{pt \notin Q\}$ *if* $\partial M \backslash \partial Q = \emptyset$.

From 8.6 one has a Hauptvermutung relative to a codimension 0 submanifold, which we leave the reader to formulate precisely.

Now suppose $Q \subset M$ is a codimension $q$ proper submanifold and $\xi/Q$ a normal block bundle. Let $HT(M,\xi)$ denote the $\Delta$-set of homotopy triangulations which are block homotopy equivalences on $E(\eta) = h^{-1}(E(\xi))$ (and hence in particular a $PL$ isomorphism on $Q_1$ = zero-section of $\eta$).

**(8.7) Corollary.** *The natural inclusion defines a homotopy equivalence*

$$HT(M,\xi) \;\simeq\; HT(M)$$

*if* $n \geq 6, q \geq 3$ *and* $\pi_1(M) = \pi_1(\partial M) = 0$.

**Proof.** Consider the diagram

$$\begin{array}{ccccc}
HT(M|E(\xi)) & & \xrightarrow{\quad\simeq 8.6\quad} & & (G/PL)^{M_0/Q} \\
\downarrow & & & & \downarrow \\
HT(M,\xi) & \xrightarrow{\text{inc.}} & HT(M) & \xrightarrow{\simeq 3.3\text{-}5} & (G/PL)^{M_0} \\
\downarrow{\scriptstyle\text{restriction}} & & & & \downarrow{\scriptstyle\text{restriction}} \\
HT(\xi) & & \xrightarrow{\quad\simeq 8.4\quad} & & (G/PL)^{Q}
\end{array}$$

The outside vertical maps are fibrations, commutativity of the top square is clear and of the bottom square (up to sign) follows from the proof of 8.5. The result now follows from the 5-lemma.

**Relative Hauptvermutung.**

We apply 8.7 to give a Hauptvermutung relative to a submanifold of codimension $\geq 3$.

**(8.8) Theorem.** *Suppose $M$ satisfies the conditions of the Hauptvermutung and $Q \subset M$ is a proper codimension $\geq 3$ submanifold. Then any homeomorphism $h : (M_1, Q_1) \longrightarrow (M, Q)$, which is a $PL$ isomorphism on $Q_1$, is homotopic mod $Q_1$ to a $PL$ isomorphism.*

**(8.9) Theorem.** *Suppose $M$ and $Q$ both satisfy the conditions of the Hauptvermutung and $Q \subset M$ is a proper codimension $\geq 3$ submanifold. Then any homeomorphism of pairs $h : (M_1, Q_1) \longrightarrow (M, Q)$ is homotopic to a $PL$ isomorphism of pairs.*

**Proofs.** In 8.8 it is easy to homotope $h$ to be a b.h.e. on some block neighborhood $\eta$ of $Q_1$ in $M$. $h \simeq PL$ isomorphism by the main theorem and it is homotopic via maps which are b.h.e.'s on $\eta$ by Corollary 8.7.

In 8.9 one first homotopes $h|_{Q_1}$ to a $PL$ isomorphism and extends to give a homotopy equivalence $h_1 : M_1 \longrightarrow Q_1$. It is again easy to make $h_1$ a b.h.e. on $\eta$ and then the proof of 8.8 works.

**The embedding theorem.**

**(8.10) Theorem.** *Suppose $f : M^n \longrightarrow Q^{n+q}$ is a simple homotopy equivalence, $M$ closed and $q \geq 3$. Then $f$ is homotopic to a $PL$ embedding.*

**(8.11) Corollary.** *Suppose $f : M^n \longrightarrow Q^{n+q}$ is $(n-q+1)$-connected, then $f$ is homotopic to a $PL$ embedding.*

**Proof.** This follows at once from 8.10 and Stallings [40].

**Proof of 8.10.** If $n+q<6$ the theorem is trivial, so assume $n+q\geq 6$. Let $g : \partial Q \longrightarrow M$ be the homotopy inverse of $f$ restricted to $\partial Q$.

**Assertion 1.** As a fibration, $g$ is fibre homotopy equivalent to the projection of a sphere block bundle $g_1 : E(\partial\eta) \longrightarrow M$.

The theorem then follows by replacing $g$ and $g_1$ by their mapping cylinders to obtain (up to homotopy type) :

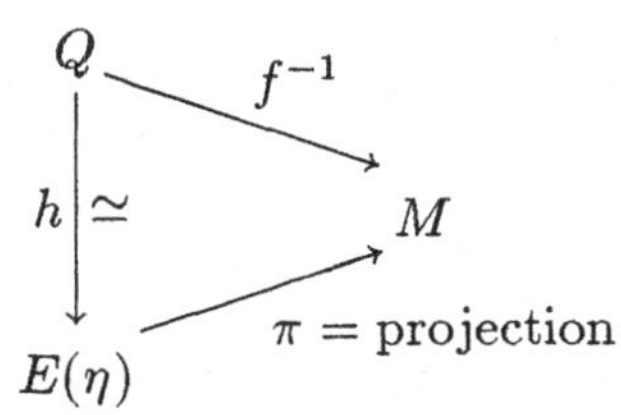


where $h$ is a homotopy equivalence $(Q, \partial Q) \longrightarrow (E(\eta), E(\partial\eta))$. But $f^{-1}$ and $\pi$ are both simple homotopy equivalences so $h$ is a simple homotopy triangulation of $E(\eta)$ and hence by 8.5 homotopic to a b.h.e. Therefore $Q$ is $PL$ isomorphic to a block bundle over $M$ and $M$ is embedded in $Q$ (by a map homotopic to $f$).

Instead of Assertion 1, we prove:

**Assertion 2.** Some large suspension (along the fibres) of $g : \partial Q \longrightarrow M$ is fibre homotopy equivalent to the projection of a sphere block bundle over $M$.

From this it follows that the fibre of $g$ suspends to a homology sphere (and being simply connected) must be a homotopy sphere. Assertion 1 then follows at once from the classifying space version of 2.1(d) which asserts that

$$\begin{array}{ccc} B\widetilde{PL}_q & \longrightarrow & B\widetilde{G}_q \simeq BG_q \\ \downarrow & & \downarrow \\ B\widetilde{PL} & \longrightarrow & B\widetilde{G} \simeq BG \end{array}$$

is a pushout for $q \geq 3$, i.e. "a spherical fibre space stably equivalent to a sphere block bundle is already equivalent to one".

Now to prove Assertion 2 we only have to notice that $f \times \mathrm{id} : M \longrightarrow Q \times I^N$ ($N$ large) is homotopic to the inclusion of the zero section of a block bundle. First shift to an embedding $f_1$, then choose a normal block bundle $\xi/f_1M$ and observe (cf. Mazur [24]) that $c\ell(Q \times I^N \backslash E(\xi))$ is an $s$-cobordism and hence a product. So we may assume $E(\xi) = Q \times I^N$, as required.

Now $f_1^{-1}|_{\partial(Q\times I^N)}$ is the projection of a sphere block bundle and the suspension along the fibres of $g$.

**Remarks.** (1) There is a similarly proved relative version of 8.10 in case $M$ and $Q$ are bounded and $f|_{\partial M}$ is an embedding in $\partial Q$. Hence using Hudson [16] one has that any two embeddings homotopic to $f$ are isotopic.
(2) 8.10 (and the above remark) reduce the embedding and knot problems to "homotopy theory" – one only has to embed up to homotopy type. The reduction of the problem to homotopy theory by Browder [3] follows easily from this one – for Browder's smooth theorems, one combines the $PL$ theorems with smoothing theory using Haefliger [12] and Rourke and Sanderson [32].

## References


[1] W. Browder, **The homotopy type of differentiable manifolds**, Proceedings of the Aarhus Symposium (1962) 42–46.
reprinted in Proceedings of 1993 Oberwolfach Conference on the Novikov Conjectures, Index Theorems and Rigidity (ed. S.Ferry, A.Ranicki and J.Rosenberg), Vol. 1, LMS Lecture Notes 226, Cambridge (1995) 97–100.

[2] W. Browder, **Manifolds with $\pi_1 = \mathbb{Z}$**, Bull. Amer. Math. Soc. 72 (1966) 238–244.

[3] W. Browder, **Embedding smooth manifolds**, Proc. I.C.M. Moscow, (1966) 712–719.

[4] W. Browder, **The Kervaire invariant of framed manifolds and its generalization**, Ann. of Math. 90 (1969) 157–186.

[5] W. Browder and M. W. Hirsch, **Surgery on piecewise linear manifolds and applications**, Bull. Amer. Math. Soc. 72 (1966) 959–964.

[6] E. H. Brown, **Cohomology theories**, Ann. of Math. 75 (1962) 467–484.

[7] E. H. Connell and J. Hollingsworth, **Geometric groups and Whitehead**

**torsion**, Trans. Amer. Math. Soc. 140 (1969) 161–181.

[8] P. E. Conner and E. E. Floyd, **Differentiable and Periodic Maps**, Ergebnisse der Mathematik 33, Springer Verlag, Berlin (1964).

[9] A. Dold, **Partitions of unity in the theory of fibrations**, Ann. of Math. 78 (1963) 223–255.

[10] H. Gluck, **Embeddings in the trivial range**, Ann. of Math. 81 (1965) 195–210.

[11] A. Haefliger, **Knotted spheres and related geometric topics**, Proc. I.C.M. Moscow (1966) 437–445.

[12] A. Haefliger, **Lissage des immersions I**, Topology 6 (1967) 221–240.

[13] A. Haefliger and V. Poenaru, **La classification des immersions combinatoires**, Inst. Hautes Études Sci. Publ. Math. 23 (1964) 75–91.

[14] A. Haefliger and C. T. C. Wall, **Piecewise linear bundles in the stable range**, Topology 4 (1965) 209–214.

[15] M. W. Hirsch, **On normal microbundles**, Topology 5 (1966) 229–240.

[16] J. F. P. Hudson, **Concordance and isotopy of *PL* embeddings**, Bull. Amer. Math. Soc. 72 (1966) 534–536.

[17] I. M. James, **On the iterated suspension**, Quart. Jour. Math. (Oxford) 5 (1954) 1–10.

[18] M. A. Kervaire, **A manifold which does not admit any differentiable structure**, Comm. Math. Helv. 70 (1960) 257–270.

[19] M. A. Kervaire and J. W. Milnor, **Groups of homotopy spheres I**, Ann. of Math. 77 (1963) 504–537.

[20] N. H. Kuiper and R. K. Lashof, **Microbundles and bundles I**, Invent. Math. 1 (1966) 1–17.

[21] R. C. Lacher, **Cell–like mappings of *ANR*'s**, Bull. Amer. Math. Soc. 74 (1968) 933–935.

[22] R. K. Lashof and M. G. Rothenberg, **Hauptvermutung for manifolds**, Conference on the topology of manifolds, Prindle, Weber and Schmidt (1968) 81–105.

[23] J. Levine, **A classification of differentiable knots**, Ann. of Math. 82 (1965) 15–50.

[24] B. Mazur, **Differential topology from the point of view of simple homotopy theory**, Inst. Hautes Études Sci. Publ. Math. 15 (1963).

[25] J. W. Milnor, **Differentiable manifolds which are homotopy spheres**, mimeo, Princeton (1959).

[26] J. W. Milnor, **Microbundles I**, Topology 3 (1964) 53–80.

[27] S. P. Novikov, **Topological invariance of rational Pontrjagin classes**, Doklady 163 (1965) 298–300. (Soviet Math. (1965) 921–923.)

[28] S. P. Novikov, **Homotopically equivalent smooth manifolds I**, Izv. Akad. Nauk. S.S.S.R. Ser. Mat. 28 (1964) 365–474. (Translations Amer. Math. Soc. 48 (1965) 271–396.)

[29] V. A. Rohlin, **New results in the theory of 4-dimensional manifolds**,

Doklady 8 (1952) 221–224.

[30] C. P. Rourke and B. J. Sanderson, **Block bundles I**, Ann. of Math. 87 (1968) 1–28.

[31] C. P. Rourke and B. J. Sanderson, **Block bundles II**, Ann. of Math. 87 (1968) 256–278.

[32] C. P. Rourke and B. J. Sanderson, **Block bundles III**, Ann. of Math. 87 (1968) 431–483.

[33] C. P. Rourke and B. J. Sanderson, **Δ-sets I**, Quart. Jour. Math. (Oxford) 22 (1971) 321–338.

[34] C. P. Rourke and B. J. Sanderson, **Δ-sets II**, Quart. Jour. Math. (Oxford) 22 (1971) 465–485.

[35] C. P. Rourke and B. J. Sanderson, **Piecewise linear topology**, Ergebnisse der Mathematik 69, Springer Verlag, Berlin (1972).

[36] C. P. Rourke and D. P. Sullivan, **On the Kervaire obstruction**, Ann. of Math. 94 (1971) 397–413.

[37] L. C. Siebenmann, **The obstruction to finding a boundary for an open manifold of dimension greater than five**, Ph.D. Thesis, Princeton (1965).

[38] M. Spivak, **On spaces satisfying Poincaré duality**, Topology 6 (1967) 77–101.

[39] J. R. Stallings, **The piecewise linear structure of euclidean space**, Proc. Cam. Phil. Soc. 58 (1962) 481–488.

[40] J. R. Stallings, **The embedding of homotopy types into manifolds**, mimeo, Princeton (1965).

[41] D. P. Sullivan, **Triangulating homotopy equivalences**, **Ph.D. Thesis** Princeton (1966).

[42] C. T. C. Wall, **An extension of results of Browder and Novikov**, Amer. Jour. Math. 88 (1966) 20–32.

[43] C. T. C. Wall, **Poincaré complexes I**, Ann. of Math. 86 (1967) 213–245.

[44] C. T. C. Wall, **Surgery On Compact Manifolds**, Academic Press, London (1970).

[45] R. E. Williamson, **Cobordism of combinatorial manifolds**, Ann. of Math. 83 (1966) 1–33.

[46] E. C. Zeeman, **Seminar on combinatorial topology**, Inst. Hautes Études Sci., Paris (1963–66).

Institute for Advanced Study
Princeton
Spring 1968

# CHAPTER III

## The Hauptvermutung according to Casson and Sullivan

By G. E. Cooke

### §1. Introduction

Chapter II contains a proof, due to Casson and Sullivan, of the Hauptvermutung for a $PL$ manifold $M$ which satisfies:

(1) $M$ is either closed of dimension at least five, or bounded of dimension at least six;
(2) each component of both $M$ and $\partial M$ is simply connected;
(3) $H^3(M;\mathbb{Z}_2) = 0$.

In this chapter we give a proof that condition (3) can be weakened to:

(3)′ $H^4(M;\mathbb{Z})$ has no 2-torsion.

This stronger result was stated in Sullivan [18]. An outline of the original proof may be found in Sullivan [17]. The argument presented here was sketched in §6 of Chapter II.

The main result is conveniently stated in terms of the obstructions to deforming a homotopy equivalence to a homeomorphism introduced by Sullivan in his thesis [16]. Assume that $h: Q \longrightarrow M$ is a topological homeomorphism. Write $M_0 = M$ if $\partial M$ is non-empty, and $M_0 = M\backslash\{\text{disc}\}$ when $\partial M$ is empty. The $i$-th obstruction to deforming $h$ to a $PL$ homeomorphism is denoted by $o^i(h)$. It lies in $H^i(M_0;\pi_i(G/PL))$. We shall prove that:

(a) $o^4(h) \in H^4(M_0;\pi_4(G/PL)) = H^4(M_0;\mathbb{Z})$ is defined (all earlier obstructions are zero) and is an element of order at most two;
(b) if $o^4(h) = 0$ then all higher obstructions vanish.

Our method is to use information about the bordism groups of $G/PL$ to yield results on the $k$-invariants of $G/PL$. The following auxiliary result, which was originally stated in Sullivan [17] and which implies that the $4i$-th $k$-invariants are of odd order for $i > 1$, is of independent interest:

**(4.4) Theorem.** *There exist classes $\ell_i \in H^{4i}(G/PL;\mathbb{Z}_{(2)})$ for each $i \geq 0$ such that if we write*

$$\mathcal{L} = \ell_0 + \ell_1 + \ell_2 + \cdots \in H^{4*}(G/PL;\mathbb{Z}_{(2)})$$

*then for any map* $f : M^{4k} \longrightarrow G/PL$ *of a smooth manifold the surgery obstruction of the map* $f$ *(see §4 of Chapter II) is given by*

$$s(f) = 8\langle L(M) \cup f^*\mathcal{L}, [M]\rangle ,$$

*where* $L(M)$ *is the Hirzebruch* $L$*-genus, and* $\mathbb{Z}_{(2)}$ *denotes the integers localized at 2.*

We reproduce a proof given by Sullivan in his thesis that the $(4i+2)$-nd $k$-invariants of $G/PL$ are zero, and explicitly calculate the 4th $k$-invariant (Theorem 4.6).

§2 contains two elementary results on principal fibrations. We follow the treatment of Spanier [14] so closely that proofs are unnecessary. In §3 various results on Postnikov systems are stated. The literature on Postnikov system is scattered. 2-stage Postnikov systems were discussed by Eilenberg and MacLane [4]; in particular, the notation $k$-invariant is due to them. Postnikov's fundamental papers appeared in 1951; see [10] for an English translation. Other basic references are Moore [8, 9]; recent treatments are in Spanier [14] and Thomas [20]. In §3 I give proofs of two well-known elementary results (3.7 and 3.8) which I was unable to find in the literature. I also quote a result of Kahn's [6] on Postnikov systems of $H$-spaces because of its general interest. In §4 the desired results on the homotopy properties of $G/PL$ are obtained and applied to the Hauptvermutung.

I am happy to acknowledge the substantial help of several people in the preparation of this chapter. I wish to thank Colin Rourke for arousing my interest in the problem and for explaining much of the needed geometry such as Sullivan's thesis. I am grateful to Greg Brumfiel for outlining Sullivan's proof of the main result to me, and for showing me how to extend Theorem 4.3 to $PL$-manifolds. And I thank Bob Stong for patiently explaining the necessary results in cobordism theory – especially Theorem 4.5.

## §2. Principal fibrations

Let $B$ be a space with base point $b_0$. Let $PB$ denote the space of paths in $B$ starting at $b_0$. The evaluation map

$$p : PB \longrightarrow B \; ; \; \lambda \longrightarrow \lambda(1)$$

is the projection of the **path-space fibration** (see Spanier [14]). The fibre of $p$ is the space of loops in $B$ based at $b_0$, which is denoted $\Omega B$. If $f : X \longrightarrow B$ is a map, the induced fibration over $X$ is called the principal $\Omega B$-fibration induced by $f$. The total space $E$ is defined by

$$E = \{(x, \lambda) \in X \times PB \mid f(x) = \lambda(1)\}$$

and the projection $E \xrightarrow{\pi} X$ is defined by $\pi(x, \lambda) = x$. ($E$ is often called the **fibre** of the map $f$.) Suppose that $x_0 \in X$ is a base point and that $f(x_0) = b_0$. Then we stipulate that $c_0 = (x_0, \lambda_0)$ is the base point of $E$, where $\lambda_0$ is the constant path. We have an inclusion $\Omega B \xrightarrow{\pi} E$ given by $j(\lambda) = (x_0, \lambda)$. Let $Y$ be a space with base point $y_0$. In the following theorem $[Y, \cdot]$ denotes the functor "homotopy-rel-base point classes of base-point preserving maps from $Y$ to $\cdot$".

**(2.1) Theorem.** *The following sequence of pointed sets is exact:*

$$[Y, \Omega B] \xrightarrow{j_*} [Y, E] \xrightarrow{\pi_*} [Y, X] \xrightarrow{f_*} [Y, B] .$$

We define an action

$$m \; : \; \Omega B \times E \longrightarrow E \; ; \; (\lambda, (x, \lambda')) \longrightarrow (x, \lambda * \lambda') \; ,$$

where $\lambda * \lambda'$ denotes, as usual, the path

$$(\lambda * \lambda')(t) \;=\; \begin{cases} \lambda(2t) & t \le \frac{1}{2} \\ \lambda'(2t-1) & t \ge \frac{1}{2} \, . \end{cases}$$

The action $m$ is consistent with the inclusion $j : \Omega B \longrightarrow E$ and the multiplication in $\Omega B$ since

$$\begin{aligned} m(\lambda, j(\lambda')) &= m(\lambda, (x_0, \lambda')) \\ &= (x_0, \lambda * \lambda') \\ &= j(\lambda * \lambda') \, . \end{aligned}$$

The map $m$ induces an action

$$m_* \; : \; [Y, \Omega B] \times [Y, E] \longrightarrow [Y, E]$$

where $[Y, \Omega B]$ inherits a group structure from the multiplication in $\Omega B$.

**(2.2) Theorem.** *If $u, v$ are elements of $[Y, E]$, then $\pi_* u = \pi_* v$ if and only if there exists $w$ in $[Y, \Omega B]$ such that*

$$v \;=\; m_*(w, u) \, .$$

## §3. Postnikov systems

Let $X$ be a topological space. A **cofiltration** of $X$ is a collection of spaces $\{X_i\}$ indexed on the non-negative integers and maps

$$f_i \; : \; X \longrightarrow X_i \quad , \quad g_i \; : \; X_i \longrightarrow X_{i-1}$$

such that the composition $g_i f_i$ is homotopic to $f_{i-1}$. A cofiltration is usually assembled in a diagram as below:

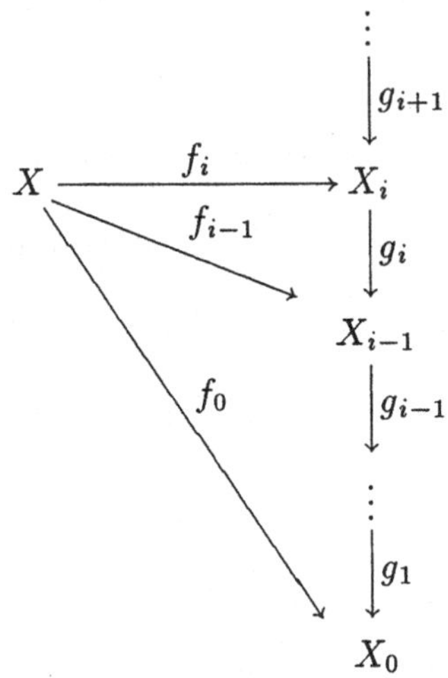


A cofiltration is called **convergent** if for any integer $N$ there is an $M$ such that for all $m > M$, $f_m : X \longrightarrow X_m$ is $N$-connected. The notion of convergent cofiltration has many applications. For example, a very general problem in topology is to determine the homotopy classes of maps of a finite complex $K$ to a space $X$ (denoted $[K, X]$). If $\{X_i\}, \{f_i\}, \{g_i\}$ is a convergent cofiltration of $X$, then the problem of calculating $[K, X]$ is split up into a finite sequence of problems: given that $[K, X_i]$ is known, calculate $[K, X_{i+1}]$. Since the cofiltration is assumed convergent, $(f_i)_* : [K, X] \longrightarrow [K, X_i]$ is a bijection for large $i$ and so the sets $[K, X_i]$ converge after a finite number of steps to a solution of the problem.

Naturally one would like to concentrate on cofiltrations where each step of the above process (calculate $[K, X_{i+1}]$ given $[K, X_i]$) is as simple as possible. Obstruction theory leads to the following requirement for simplifying the problem: that each map $g_i : X_i \longrightarrow X_{i-1}$ be a fibration with fibre an Eilenberg-MacLane space. The precise definition is given in the following way.

First note that the space of loops in a $K(\pi, n+1)$ is a $K(\pi, n)$, and so it makes sense to speak of a principal $K(\pi, n)$-fibration. Such a fibration is induced by a map of the base space into $K(\pi, n+1)$.

**(3.1) Definition.** A **Postnikov system** for a path-connected space $X$ is a convergent cofiltration

$$\{X_i\} \quad , \quad \{f_i : X \longrightarrow X_i\} \quad , \quad \{g_i : X_i \longrightarrow X_{i-1}\}$$

of $X$ such that $X_0$ is contractible, each $X_i$ has the homotopy type of a $CW$ complex, and $g_i : X_i \longrightarrow X_{i-1}$ $(i > 0)$ is a principal $K(\pi, i)$-fibration for some $\pi$.

Suppose that $X$ is a path-connected space, and that $\{X_i\}$, $\{f_i : X \longrightarrow X_i\}$,

$\{g_i : X_i \longrightarrow X_{i-1}\}$ is a Postnikov system for $X$. Then it follows that:

(a) For each $i$, the map $f_i : X \longrightarrow X_i$ induces isomorphisms of homotopy groups in dimensions $\leq i$, and $\pi_j(X_i) = 0$ for $j > i$;

(b) for each $i > 0$, the map $g_i : X_i \longrightarrow X_{i-1}$ is actually a principal $K(\pi_i(X), i)$ fibration, induced by a map

$$k^i : X_{i-1} \longrightarrow K(\pi_i(X), i+1) .$$

The class $k^i \in H^{i+1}(X_{i-1}; \pi_i(X))$ is called the $i$-th $k$-invariant of the given Postnikov system. Furthermore, the inclusion of the fibre of $g_i$, $j : K(\pi_i(X), i) \longrightarrow X_i$, induces a homology homomorphism equivalent to the Hurewicz homomorphism $h_i$ in the space $X$, in that the diagram below is commutative:

$$\begin{array}{ccc} H_i(K(\pi_i(X), i)) & \xrightarrow{j_*} & H_i(X_i) \\ \simeq \downarrow & & \uparrow (f_i)_* \\ \pi_i(X) & \xrightarrow{h_i} & H_i(X) \end{array}$$

(c) $X$ is a simple space; that is to say, $\pi_1(X)$ acts trivially on $\pi_n(X)$ for each $n$, or equivalently, any map of the wedge $S^1 \vee S^n \longrightarrow X$ extends over the product $S^1 \times S^n$. In particular, $\pi_1(X)$ is abelian.

On the other hand, if $X$ is simple then there exists a Postnikov system for $X$. See Spanier [14, Corollary 8.3.1, p.444]. For a discussion of the uniqueness of Postnikov systems, see Barcus and Meyer [1].

The following facts about Postnikov systems are presented without proofs except in cases where I do not know of an appropriate reference:

**(3.2) Maps of Postnikov systems.** Let $X$ and $X'$ be spaces, with Postnikov systems $\{X_i, f_i, g_i\}$, $\{X_i', f_i', g_i'\}$. If $h : X \longrightarrow X'$ is a map, then there exists a map of Postnikov systems consistent with the map $h$; that is, there is a collection of maps $\{h_i : X_i \longrightarrow X_i'\}$ such that $h_i f_i \simeq f_i' h$ and $h_{i-1} g_i \simeq g_i' h_i$ for all $i$. (See Kahn [6].)

**(3.3) Cohomology suspension.** If $B$ is a space with base point $*$, and $PB \xrightarrow{p} B$ is the path space fibration, then for $i > 1$ the composition

$$H^i(B;G) \xleftarrow{\approx} H^i(B,*;G) \xleftarrow{p^*} H^i(PB, \Omega B; G) \underset{\approx}{\xleftarrow{\delta}} H^{i-1}(\Omega B; G) \longrightarrow H^{i-1}(\overline{\Omega} B; G)$$

(with $G$ an arbitrary coefficient group) is called the **cohomology suspension** and is denoted by $\sigma : H^i(B;G) \longrightarrow H^{i-1}(\overline{\Omega} B; G)$. For $i \leq 1$ $\sigma$ is set equal to zero.

**(3.4) Postnikov system of a loop space.** (See Suzuki [19] for a study of the case of 2 non-vanishing homotopy groups.) Let $X$ be a path-connected space with base point. Let $\overline{\Omega}X$ denote the component of the loop space of $X$ consisting of those loops which are homotopic to a constant. If $\{X_i\}, \{f_i : X \longrightarrow X_i\}$ $\{g_i : X_i \longrightarrow X_{i-1}\}$ is a Postnikov system for $X$, then a Postnikov system for $\overline{\Omega}X$ is obtained by applying the loop space functor. That is, set

$$\begin{aligned} Y_i &= \overline{\Omega}X_{i+1} , \\ f_i' &= \Omega f_{i+1} : \overline{\Omega}X \longrightarrow \overline{\Omega}X_{i+1} , \\ g_i' &= \Omega g_{i+1} : \overline{\Omega}X_{i+1} \longrightarrow \overline{\Omega}X_i \end{aligned}$$

and then $\{Y_i, f_i', g_i'\}$ is a Postnikov system for $\overline{\Omega}X$. The $k$-invariants of this Postnikov system for $\overline{\Omega}X$ are just the cohomology suspensions of the $k$-invariants of the Postnikov system $\{X_i, f_i, g_i\}$.

**(3.5) Definition.** For any space $Y$, the set $[Y, \overline{\Omega}B]$ inherits a group structure from the multiplication on $\overline{\Omega}B$. We shall often use the fact that if $u \in H^i(B; G)$ then $\sigma u \in H^{i-1}(\overline{\Omega}B; G)$ is **primitive** with respect to the multiplication on $\overline{\Omega}B$; this means that for any space $Y$ and for any two maps $f, g \in [Y, \overline{\Omega}B]$,

$$(f \cdot g)^* \sigma u = f^* \sigma u + g^* \sigma u ,$$

where $f \cdot g$ denotes the product of $f$ and $g$. (See Whitehead [22].)

**(3.6) Postnikov system of an $H$-space.** Let $X$ be an $H$-space. Then $X$ is equipped with a multiplication $h : X \times X \longrightarrow X$ such that the base point acts as a unit. If $X$ and $Y$ are $H$-spaces, then a map $f : X \longrightarrow Y$ is called an $H$-map if $fh_X \simeq h_Y(f \times f)$. It is proved by Kahn [6] that if $X$ is an $H$-space and $\{X_i, f_i, g_i\}$ is a Postnikov system for $X$, then each $X_i$ can be given an $H$-space structure in such a way that:

(a) for all $i$, $f_i$ and $g_i$ are $H$-maps,

(b) for all $i$, the $k$-invariant $k^i \in H^{i+1}(X_{i-1}; \pi_i(X))$ is primitive with respect to the multiplication on $X_{i-1}$.

**(3.7) Vanishing of $k$-invariant.** If $X$ is a space and

$$\begin{array}{c} X_i \\ \Big\downarrow g_i \\ X_{i-1} \xrightarrow{\ k^i\ } K(\pi_i(X), i+1) \end{array}$$

is the $i$-th stage of a Postnikov system for $X$, then $k^i = 0$ if and only if the Hurewicz map $h : \pi_i(X) \longrightarrow H_i(X)$ is a monomorphism onto a direct summand.

**Proof.** Serre [12] constructs for any fibre space such that the fundamental group of the base acts trivially on the homology of the fibre an exact sequence of homology groups. In the case of the fibration $g_i : X_i \longrightarrow X_{i-1}$, with coefficient group $\pi_i(X)$, we obtain

$$\dots \longrightarrow H^i(X_i;\pi_i(X)) \xrightarrow{j^*} H^i(K(\pi_i(X),i);\pi_i(X)) \xrightarrow{\tau} H^{i+1}(X_{i-1};\pi_i(X))\,.$$

Here $j : K(\pi_i(X),i) \longrightarrow X_i$ is the inclusion of the fibre and $\tau$ is the transgression. The fundamental group of $X_{i-1}$ acts trivially on the homology of the fibre because the fibre space is induced from the path-space fibration over $K(\pi_i(X), i+1)$. The sequence is exact in the range needed even if $X_{i-1}$ is not simply-connected, as a simple argument using the Serre spectral sequence will show.

Let $\imath \in H^i(K(\pi_i(X),i);\pi_i(X))$ denote the fundamental class. The natural isomorphism

$$H^i(K(\pi_i(X),i);\pi_i(X)) \;\cong\; \mathrm{Hom}(\pi_i(X),\pi_i(X))$$

sends $\imath$ to the identity map. Consider the square below:

$$\begin{array}{ccc} X_i & \longrightarrow & PK(\pi_i(X),i+1) \\ {\scriptstyle K(\pi_i(X),i)}\big\downarrow & & \big\downarrow{\scriptstyle K(\pi_i(X),i)} \\ X_{i-1} & \longrightarrow & K(\pi_i(X),i+1) \end{array}$$

In the path-space fibration the fundamental classes of the fibre and base space correspond under transgression; the $k$-invariant $k^i \in H^{i+1}(X_{i-1};\pi_i(X))$ is by definition the pull-back of the fundamental class of the base space $K(\pi_i(X),i+1)$. It follows that $\tau(\imath) = k^i$.

First suppose $k^i = 0$. Then by Serre's exact sequence there is a class $x \in H^i(X_i;\pi_i(X))$ such that $j^*x = \imath$. The action of $x$ on the homology of $X_i$ gives a map such that the diagram below is commutative:

$$\begin{array}{ccc} & & \pi_i(X) \\ & \nearrow^{x} & \\ H_i(X_i) & & \big\downarrow\cong \\ & \nwarrow_{j_*} & \\ & & H_i(K(\pi_i(X),i)) \end{array}$$

But $j_*$ is essentially the Hurewicz homomorphism $h : \pi_i(X) \longrightarrow H_i(X)$ by remark (b) above, so $x$ gives a splitting map and $h$ is a monomorphism onto a direct summand.

Now assume that $h$ is a monomorphism onto a direct summand. Then so is $j_*$, and we may choose a splitting map

$$p \;:\; H_i(X_i) \longrightarrow \pi_i(X)$$

such that the diagram above is commutative with $p$ in place of $x$. The universal coefficient theorem implies that there is a class $x \in H^i(X_i;\pi_i(X))$ whose action on

$H_i(X_i)$ is the map $p$. It follows that $j^*x = \imath$, since $K(\pi_i(X), i)$ has no homology in dimension $i-1$. Thus $k^i = 0$ and the proof is complete.

**(3.8) Order of $k$-invariant**. Let $X$ be a space such that $\pi_i(X) = \mathbb{Z}$ for some $i$. Then the $i$-th $k$-invariant $k^i$ in any Postnikov system for $X$ is of finite order if and only if there is a cohomology class in $H^i(X)$ which takes a non-zero value on the generator of $\pi_i(X)$. The order of $k^i$ is equal to the least positive integer $d$ such that there is a cohomology class in $H^i(X)$ which takes the value $d$ on the generator of $\pi_i(X)$.

**Proof.** Let

$$\begin{array}{ccc} X_i & & \\ \Big\downarrow{\scriptstyle K(\mathbb{Z},i)} & & \\ X_{i-1} & \longrightarrow & K(\mathbb{Z}, i+1) \end{array}$$

be the $i$-th stage of a Postnikov system for $X$. We have as in 3.7 an exact sequence

$$\cdots \longrightarrow H^i(X_i) \xrightarrow{j^*} H^i(K(\mathbb{Z},i)) \xrightarrow{\tau} H^{i+1}(X_{i-1})$$

and the fundamental class $\imath \in H^i(K(\mathbb{Z},i))$ transgresses to $k^i$. It follows from remark (b) above that after identifying $H^i(X)$ with $H^i(X_i)$ the map $j^*$ can be regarded as evaluation of $H^i(X)$ on $\pi_i(X)$. In other words

$$\begin{array}{ccc} H^i(X_i) & \xrightarrow{j^*} & H^i(K(\mathbb{Z},i)) \\ \Big\downarrow{\scriptstyle (f_i)^*} & & \Big\downarrow{\scriptstyle \cong} \\ H^i(X) & \xrightarrow{\text{eval.}} & \mathrm{Hom}(\pi_i(X), \mathbb{Z}) \end{array}$$

is commutative. Thus to prove the first part of (3.8) we have

$$\begin{aligned} k^i \text{ is of infinite order} &\iff \tau \text{ is a monomorphism} \\ &\iff j^* = 0 \\ &\iff \text{every cohomology class in } H^i(X) \\ &\qquad \text{takes the value 0 on } \pi_i(X)\,. \end{aligned}$$

To prove the second statement of (3.8) we have

$$\begin{aligned} \text{the order of } k^i \text{ divides } d &\iff j^*x = d\imath \text{ for some } x \in H^i(X_i) \\ &\iff \text{some cohomology class in} \\ &\qquad H^i(X) \text{ takes the value } d \text{ on} \\ &\qquad \text{the generator of } \pi_i(X)\,. \end{aligned}$$

## §4. Application to $G/PL$ and the Hauptvermutung

Recall that in Chapter II it is proved that the periodicity map

$$\mu : G/PL \longrightarrow \overline{\Omega}^{4n}(G/PL)$$

has fibre an Eilenberg-MacLane space $K(\mathbb{Z}_2, 3)$. Furthermore, the composition

$$TOP/PL \xrightarrow{\ i\ } G/PL \xrightarrow{\ \mu\ } \overline{\Omega}^{4n}(G/PL)$$

is null-homotopic. Now let $M$ and $Q$ be $PL$ manifolds, and assume that

$$h : Q \longrightarrow M$$

is a topological homeomorphism. Associated to $h$ is a map

$$q_h : M_0 \longrightarrow G/PL$$

(where $M_0 = M\backslash\{\text{disc}\}$ if $\partial M = \emptyset$, $M_0 = M$ if $\partial M \neq \emptyset$) and, under certain conditions on $M$, $h$ is homotopic to a $PL$ homeomorphism if and only if $q_h$ is homotopic to a constant. (For example, it is enough to assume:

(1) $M$ and $Q$ are closed of dimension $\geq 5$ or bounded of dim $\geq 6$

(2) $\pi_1(M) = \pi_1(\partial M) = 0$ .)

In this section we study the question of whether the map $q_h$ is null-homotopic. By Chapter II the map $q_h$ factors through $TOP/PL$:

$$\begin{array}{ccc} & TOP/PL & \\ \nearrow & & \searrow^{i} \\ M_0 & \xrightarrow{\qquad q_h \qquad} & G/PL \end{array}$$

and so the composition $\mu q_h : M_0 \longrightarrow \overline{\Omega}^{4n} G/PL$ is null-homotopic. By Theorems 2.1 and 2.2, this means that

(a) $q_h$ lifts to a map into the total space of the $\Omega\overline{\Omega}^{4n}(G/PL)$ fibration induced by $\mu$. This total space is just the fibre of the map $\mu$ and so we denote it by $K(\mathbb{Z}_2, 3)$,

(b) different liftings of $q_h$ are related via the action of $\Omega\overline{\Omega}^{4n}(G/PL)$ on $K(\mathbb{Z}_2, 3)$.

Now any lifting of $q_h$ to a map into $K(\mathbb{Z}_2, 3)$ defines a cohomology class in $H^3(M_0; \mathbb{Z}_2)$. We shall prove:

**(4.1) Theorem.** *The collection of cohomology classes defined by liftings of $q_h$ is a coset of the subgroup of mod 2 reductions of integral classes in $M_0$ and so determines an element $V_h \in H^4(M_0; \mathbb{Z})$ of order 2. The map $q_h$ is null-homotopic if and only if $V_h = 0$.*

The theorem above can be restated in terms of the obstruction theory defined by Sullivan [16]. Let

$$o^i(q_h) \in H^i(M_0; \pi_i(G/PL))$$

denote the $i$-th obstruction to deforming $q_h$ to a constant. The theorem just stated implies that in our case (where $h$ is a topological homeomorphism):

(a) $o^2(q_h) = 0$,
(b) $o^4(q_h) \in H^4(M_0; \mathbb{Z})$ is equal to $V_h$ and is an element of order 2,
(c) if $o^4(q_h) = 0$ then all the higher obstructions vanish.

We begin with a study of the $k$-invariants of $G/PL$. Let

$$\{X_i\}\ ,\ \{f_i : G/PL \longrightarrow X_i\}\ ,\ \{g_i : X_i \longrightarrow X_{i-1}\}$$

be a Postnikov system for $G/PL$. Let

$$x^i \in H^{i+1}(X_{i-1}; \pi_i(G/PL))$$

denote the $i$-th $k$-invariant. Recall that $\pi_{4i+2}(G/PL) = \mathbb{Z}_2$, $\pi_{4i}(G/PL) = \mathbb{Z}$, and the odd groups are zero.

**(4.2) Theorem.** *For all $i$, $x^{4i+2} = 0$.*

**Proof.** This theorem was proved in Sullivan's thesis [16], and we reproduce the proof here. By 3.7 it is sufficient to show that the Hurewicz homomorphism

$$h\ :\ \mathbb{Z}_2\ =\ \pi_{4i+2}(G/PL) \longrightarrow H_{4i+2}(G/PL)$$

is a monomorphism onto a direct summand. But that is true if and only if the mod 2 Hurewicz homomorphism

$$h_2\ :\ \mathbb{Z}_2\ =\ \pi_{4i+2}(G/PL) \longrightarrow H_{4i+2}(G/PL) \xrightarrow{\text{mod 2 reduction}} H_{4i+2}(G/PL; \mathbb{Z}_2)$$

is a monomorphism.

Consider the following diagram:

$$\begin{array}{ccccc}
 & & \Omega_{4i+2}(G/PL) & & \\
 & & \downarrow \times 2 & & \\
 & \overset{h_0}{\nearrow} & \Omega_{4i+2}(G/PL) & \longrightarrow & H_{4i+2}(G/PL) \\
\pi_{4i+2}(G/PL) & & \downarrow & & \downarrow \\
 & \underset{h_1}{\searrow} & \mathfrak{N}_{4i+2}(G/PL) & \longrightarrow & H_{4i+2}(G/PL; \mathbb{Z}_2)
\end{array}$$

Here $\Omega$ and $\mathfrak{N}$ denote oriented and unoriented smooth bordism respectively. The column in the middle is exact by a result of Conner and Floyd [3]. The surgery obstruction gives a splitting

$$\pi_{4i+2}(G/PL) \xrightarrow[\quad]{\dashleftarrow} \Omega_{4i+2}(G/PL)$$

and so $h_0$ is onto a direct summand $\mathbb{Z}_2$. By the exactness of the middle column, $h_1$ is non-zero. Thus the generator $f : S^{4i+2} \longrightarrow G/PL$ of $\pi_{4i+2}$ does not bound a singular manifold in $G/PL$. According to Theorem 17.2 of [3], at least one of the Whitney numbers associated to the singular manifold $[S^{4i+2}, f]_2$ must be non-zero. Since all of the Stiefel-Whitney classes of $S^{4i+2}$ vanish (except $w_0$ which is 1), this implies that $f_*[S^{4i+2}]$ is non-zero in $\mathbb{Z}_2$-homology. Thus the mod 2 Hurewicz homomorphism is a monomorphism and Theorem 4.2 is proved.

**(4.3) Theorem.** *For all $i > 1$, $x^{4i} \in H^{4i+1}(X_{4i-1}; \mathbb{Z})$ is of odd order.*

**Proof.** By 3.8 it is sufficient to find a cohomology class in $H^{4i}(G/PL)$ which takes an odd value on the generator of $\pi_{4i}(G/PL)$.

We shall prove that such cohomology classes exist as follows: we construct, for each $i$, a cohomology class

$$\ell_i \in H^{4i}(G/PL; \mathbb{Z}_{(2)}) ,$$

where $\mathbb{Z}_{(2)}$ = integers localized at 2 = the ring of rationals with odd denominators. The classes $\ell_i$ shall be constructed so that $\ell_i$ takes the value 1 on the generator of $\pi_{4i}(G/PL)$ for $i > 1$. Since the homology of $G/PL$ is finitely generated, a sufficiently large odd multiple of $\ell_i$ is then the reduction of an integral class which takes an odd value on the generator of $\pi_{4i}(G/PL)$.

**(4.4) Theorem.** *There exist classes $\ell_i \in H^{4i}(G/PL; \mathbb{Z}_{(2)})$ for each $i \geq 0$ such that if we write*

$$\mathcal{L} = \ell_0 + \ell_1 + \ell_2 + \cdots \in H^{4*}(G/PL; \mathbb{Z}_{(2)})$$

*then for any map $f : M^{4k} \longrightarrow G/PL$ of a smooth manifold the surgery obstruction of the map $f$ (see Chapter II) is given by*

$$s(f) = 8\langle L(M) \cup f^*\mathcal{L}, [M]\rangle \tag{1}$$

*where $L(M)$ is the L-genus of Hirzebruch [5, II§8] applied to the Pontrjagin classes of $M$.*

Now if $\alpha_i \in \pi_{4i}(G/PL)$ is a generator then the surgery obstruction of the map $\alpha_i : S^{4i} \longrightarrow G/PL$ is 16 if $i = 1$ and 8 if $i > 1$. Since the Pontrjagin classes of $S^{4i}$ are trivial, Theorem 4.4 implies that

$$\langle \ell_i, \alpha_i \rangle = \begin{cases} 2 & \text{if } i = 1 \\ 1 & \text{if } i > 1 \end{cases}$$

and so Theorem 4.3 follows.

We now prove Theorem 4.1, postponing the proof of Theorem 4.4. Consider the diagram below

$$\begin{array}{ccc} & & K(\mathbb{Z}_2,3) \\ & \nearrow & \Big\downarrow \pi \\ M_0 & \xrightarrow{q_h} & G/PL \end{array}$$

Let $\imath_3 \in H^3(K(\mathbb{Z}_2,3);\mathbb{Z}_2)$ be the fundamental class. Choose a lifting $f_1 : M_0 \longrightarrow K(\mathbb{Z}_2,3)$. We prove first that given any other lifting $f_2 : M_0 \longrightarrow K(\mathbb{Z}_2,3)$ of $q_h$, we have

(a) $f_2^*\imath_3 - f_1^*\imath_3$ = reduction of an integral cohomology class.

**Proof of (a).** By Theorem 2.2, there is a map $g : M_0 \longrightarrow \Omega\overline{\Omega}^{4n}(G/PL)$ such that the composition

$$\begin{array}{ccc} M_0 & \xrightarrow{g \times f_1} & \Omega\overline{\Omega}^{4n}(G/PL) \times K(\mathbb{Z}_2,3) \\ & & \Big\downarrow m \\ & & K(\mathbb{Z}_2,3) \end{array}$$

is homotopic to $f_2$. We have

$$m^*\imath_3 \;=\; j^*\imath_3 \times 1 + 1 \times \imath_3$$

(where $j : \Omega\overline{\Omega}^{4n}(G/PL) \longrightarrow K(\mathbb{Z}_2,3)$ is the inclusion of the fibre) for dimension reasons. We evaluate $j^*\imath_3$. Since

$$\pi_1(\Omega\overline{\Omega}^{4n}(G/PL)) \;=\; \mathbb{Z}_2 \;, \quad \pi_3(\Omega\overline{\Omega}^{4n}(G/PL)) \;=\; \mathbb{Z}$$

and the even groups are zero, a Postnikov system for $\Omega\overline{\Omega}^{4n}(G/PL)$ looks like

$$\begin{array}{ccc} \Omega\overline{\Omega}^{4n}(G/PL) & \longrightarrow & E_3 \\ & \searrow & \Big\downarrow K(\mathbb{Z},3) \\ & & K(\mathbb{Z}_2,1) \end{array}$$

in low dimensions. The $k$-invariant $k^3 \in H^4(K(\mathbb{Z}_2,1);\mathbb{Z})$ is the $(4n+1)$-st suspension of the $k$-invariant $x^{4n+4}$ of $G/PL$, by 3.4. Since $x^{4n+4}$ is of odd order, so is $k^3$ (here we assume $n \geq 1$ and apply Theorem 4.3) and since $k^3$ lies in a 2-primary group it must be zero. Thus $E_3$ is a product. Hence

$$H^3(E_3;\mathbb{Z}_2) \;\cong\; \mathbb{Z}_2 \oplus \mathbb{Z}_2 \;,$$

generated by $\imath_1^3$ and $\bar{\imath}$, where $\imath_1 \in H^1(K(\mathbb{Z}_2,1);\mathbb{Z}_2)$ is the fundamental class and $\bar{\imath} \in H^3(K(\mathbb{Z},3);\mathbb{Z}_2)$ is the mod 2 reduction of the fundamental class $\imath \in H^3(K(\mathbb{Z},3);\mathbb{Z})$. Thus we may write

$$j^*\imath_3 \;=\; a\imath_1^3 + b\bar{\imath} \quad (a, b \in \mathbb{Z}_2)\ .$$

Now $\pi_3(\Omega\overline{\Omega}^{4n}(G/PL)) \cong \mathbb{Z}$ maps onto $\pi_3(K(\mathbb{Z}_2,3)) \cong \mathbb{Z}_2$ since $\pi_3(G/PL) = 0$. It follows that $b = 1$. To evaluate $a$, we consider the fibre of the map $j$, which has the homotopy type of $\Omega(G/PL)$. We obtain a sequence of spaces:

$$\Omega(G/PL) \xrightarrow{\;j'\;} \Omega\overline{\Omega}^{4n}(G/PL) \xrightarrow{\;j\;} K(\mathbb{Z}_2,3) \xrightarrow{\;\pi\;} G/PL\ .$$

We shall show that $(j')^*\imath_1^3 \neq 0$, $(j')^*\bar{\imath} = 0$. Then, since $jj' \simeq *$, it follows that $a = 0$. Let

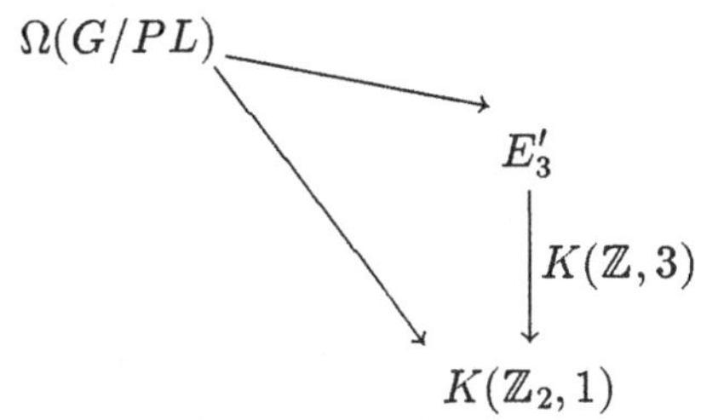


be a section of a Postnikov system for $\Omega(G/PL)$ obtained by looping the corresponding section of a Postnikov system for $G/PL$. The $k$-invariant $k^3 \in H^4(K(\mathbb{Z}_2, 1);\mathbb{Z})$ is then the suspension $\sigma x^4$ of the $k$-invariant $x^4 \in H^5(K(\mathbb{Z}_2,2);\mathbb{Z})$ for $G/PL$. We shall prove later (Theorem 4.6) that $x^4 = \delta Sq^2\imath_2$, where $\imath_2$ is the fundamental class and $\delta$ is the Bockstein operation associated to the coefficient sequence

$$\mathbb{Z} \xrightarrow{\;\times 2\;} \mathbb{Z} \longrightarrow \mathbb{Z}_2\ .$$

Since $\delta$ and $Sq^2$ commute with suspension,

$$k^3 \;=\; \sigma x^4 \;=\; \delta Sq^2\imath_1 \;=\; 0\ .$$

Thus $E_3'$ is a product. Now associated to the map $j$ is a map $E_3' \longrightarrow E_3$ by 3.2. This map multiples by 2 in $\pi_3$ and is an isomorphism on $\pi_1$, so a calculation gives

$$(j')^*\imath_1 \;=\; \imath_1\ ,\quad (j')^*\imath_1^3 \neq 0\ ,\quad (j')^*\bar{\imath} \;=\; 0\ .$$

This completes the proof that $a = 0$, and we may write

$$m^*\imath_3 \;=\; \bar{\imath}\times 1 + 1\times \imath_3\ .$$

Then since $f_2 \simeq m(g\times f_1)$ we have

$$\begin{aligned} f_2^*\imath_3 \;&=\; (g\times f_1)^*(\bar{\imath}\times 1 + 1\times\imath_3) \\ &=\; g^*\bar{\imath} + f_1^*\imath_3\ . \end{aligned}$$

The difference $f_2^*\imath_3 - f_1^*\imath_3$ is thus the reduction of the integral class $g^*\imath$, and (a) is proved.

We complete the proof of Theorem 4.1 by showing that given any lifting $f_1$ of $q_h$ and any class $u \in H^3(M_0;\mathbb{Z}_2)$ such that $u = \overline{v}$, $v \in H^3(M_0;\mathbb{Z})$, then

(b) there is a lifting $f_2$ of $q_h$ such that

$$f_2^*\imath_3 - f_1^*\imath_3 \;=\; u\ .$$

**Proof of (b).** Since $m^*\imath_3 = \overline{\imath} \otimes 1 + 1 \otimes \imath_3$ we need only find a map

$$g\ :\ M_0 \longrightarrow \Omega\overline{\Omega}^{4n}(G/PL)$$

such that $g^*\imath = (2c+1)v$ for some integer $c$. For then the map $f_2 = m(g \times f_1)$ satisfies

$$\begin{aligned} f_2^*\imath_3 &= g^*\overline{\imath} + f_1^*\imath_3 \\ &= (2c+1)\overline{v} + f_1^*\imath_3 \\ &= u + f_1^*\imath_3\ . \end{aligned}$$

**Construction of the map $g$.** Let $\{E_i\}$ denote the stages of a Postnikov system for $\Omega\overline{\Omega}^{4n}(G/PL)$. It was shown above that $E_3$ is a product, and so there exists a map

$$g_3\ :\ M_0 \longrightarrow E_3$$

such that $g_3^*\imath = v$. Now each $E_i$ is assumed to be a loop space, and so for any $i$, $[M_0, E_i]$ is a group. The map $g_3$ is constructed by lifting odd multiples of $g_3$ to successively higher stages $E_i$. Suppose we have obtained a map $g_i : M_0 \longrightarrow E_i$. The obstruction to lifting $g_i$

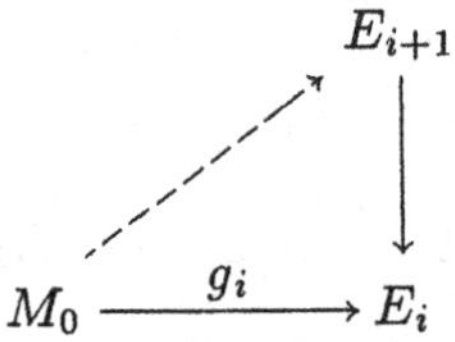

is equal to $g_i^*k^i$, where $k^i$ is the $i$-th $k$-invariant. Now the $k$-invariants are either zero or of odd order. In a case where $k^i = 0$, there is no obstruction and $g_i$ lifts. If $k^i \neq 0$ and is of odd order $2d+1$, then the map $(2d+1)g_i$ obtained by multiplying $g_i$ with itself $(2d+1)$ times in the group $[M_0, E_i]$ satisfies

$$((2d+1)g_i)^*k^i \;=\; (2d+1)g_i^*k^i \;=\; 0$$

since $k^i$ is primitive (see 3.5). Since $M_0$ is finite dimensional the obstructions vanish after a finite number of iterations of this procedure. It follows that an odd multiple of $g_3$, say $(2c+1)g_3$, lifts to a map

$$g\ :\ M_0 \longrightarrow \Omega\overline{\Omega}^{4n}(G/PL)$$

Now the class $\imath \in H^3(\Omega\overline{\Omega}^{4n}(G/PL))$ is primitive. This is true because it is actually a suspension; we argued previously that the third $k$-invariant $k^3 \in H^4(K(\mathbb{Z}_2,1);\mathbb{Z})$

is trivial because it is of odd order and 2-primary. The same is true of the fourth $k$-invariant in $H^5(K(\mathbb{Z}_2,2);\mathbb{Z})$ of $\overline{\Omega}^{4n}(G/PL)$. Thus the fourth stage of a Postnikov system for $\overline{\Omega}^{4n}(G/PL)$ splits as a product $K(\mathbb{Z}_2,2)\times K(\mathbb{Z},4)$, and the fundamental class of $K(\mathbb{Z},4)$ suspends to $\imath \in H^3(E_3)$. Since $\imath$ is primitive we have

$$g^*\imath \;=\; (2c+1)g_3^*\imath \;=\; (2c+1)v$$

and the proof of (b) is complete.

**Proof of Theorem 4.4.** We first recall some results on the smooth oriented bordism groups $\Omega_*(X)$ of a space $X$. The reader is referred to Conner and Floyd [3] or Stong [15, Chapter IX] for definitions.

Let $MSO_k$ denote the Thom complex of the universal oriented $k$-plane bundle over $BSO_k$. The spectrum $MSO = \{MSO_k\}$ classifies the bordism groups of a space $X$ in that

$$\Omega_*(X) \;=\; \pi_*(X_+ \wedge MSO)\ ,$$

with $X_+ = X \cup \{\text{pt.}\}$. Let $\mathbb{K}(\mathbb{Z},0)$ denote the Eilenberg-MacLane spectrum with $k$-th space $K(\mathbb{Z},k)$. For any connected space $X$ we have

$$\pi_*(X_+ \wedge \mathbb{K}(\mathbb{Z},0)) \;=\; H_*(X)\ .$$

The Thom class $U \in H^0(MSO)$ induces a map $MSO \longrightarrow \mathbb{K}(\mathbb{Z},0)$ which on any space $X$ yields the Hurewicz homomorphism $h : \Omega_*(X) \longrightarrow H_*(X)$. Now $\Omega_*(X)$ is an $\Omega_*(pt)$-module. An element in $\Omega_*(X)$ is **decomposable** if it is a linear combination of elements of the form

$$N_1 \times N_2 \xrightarrow{\;p_2\;} N_2 \xrightarrow{\;g\;} X$$

where $\dim N_1 > 0$. We need the following result, writing

$$\underline{G} \;=\; G/\text{torsion} \otimes \mathbb{Z}_{(2)}$$

for any group $G$.

**(4.5) Theorem.** *For any space $X$ the Hurewicz homomorphism $h : \Omega_*(X) \longrightarrow H_*(X)$ induces an epimorphism*

$$\underline{h} \;:\; \underline{\Omega_*(X)} \longrightarrow \underline{H_*(X)}$$

*with kernel generated by decomposables.*

**Proof.** According to Stong [15, p.209] the Hurewicz homomorphism in $MSO$ induces a monomorphic map of graded rings

$$h \;:\; \pi_*(MSO)/\text{torsion} \longrightarrow H_*(MSO)/\text{torsion}$$

with finite odd order cokernel in each dimension, so that

$$\underline{h} \;:\; \underline{\pi_*(MSO)} \longrightarrow \underline{H_*(MSO)}$$

is an isomorphism of graded rings. For any space $X$ we have a commutative diagram:

$$
\begin{array}{ccc}
\underline{\Omega_*(X)} & \xrightarrow{\quad h \quad} & \underline{H_*(X)} \\
\downarrow \approx & & \\
\underline{\pi_*(X_+ \wedge MSO)} & & \\
\downarrow \approx & & \downarrow \approx \\
\underline{H_*(X_+ \wedge MSO)} & & \\
\downarrow \approx & & \\
\underline{H_*(X)} \otimes \underline{H_*(MSO)} & \xrightarrow{\quad 1 \otimes U \quad} & \underline{H_*(X)} \otimes \underline{H_*(\mathbb{K}(\mathbb{Z},0))}
\end{array} .
$$

The vertical maps on the left are isomorphisms of $\underline{\pi_*(MSO)}$ (or $\underline{H_*(MSO)}$)-modules. The kernel of $1 \otimes U$ consists of decomposables, and so the same is true of the kernel of $h$.

The classes $\ell_i \in H^{4i}(G/PL; \mathbb{Z}_{(2)})$ are constructed inductively. Set $\ell_0 = 0$. Then the conclusion (1) of Theorem 4.4 holds for manifolds of dimension zero. Suppose that $\ell_0, \cdots, \ell_{i-1}$ have been defined in such a way that the conclusion of Theorem 4.4 holds for manifolds of dimension $4k, k < i$. We define the cohomology class $\ell_i$ as follows. The formula (1) forces the action of $\ell_i$ on the $4i$-th bordism group of $G/PL$; $\ell_i$ must map $\Omega_{4i}(G/PL)$ to $\mathbb{Z}_{(2)}$ by the homomorphism $\ell'$ which is defined by

$$\ell'[M^{4i}, f] \;=\; \frac{s(f)}{8} - \langle L(M) \cup f^* \sum_{j<i} \ell_j, [M] \rangle$$

for any $[M^{4i}, f] \in \Omega_{4i}(G/PL)$. The values taken by $\ell'$ lie in $\mathbb{Z}_{(2)}$ because the Hirzebruch polynomials have coefficients in $\mathbb{Z}_{(2)}$. Now suppose that $[M^{4i}, f]$ is a boundary. Then there is a smooth manifold $W^{4i+1}$ with boundary $\partial W^{4i+1} = M^{4i}$ and a map $F : W^{4i+1} \longrightarrow G/PL$ such that $F|_{\partial W} = f$. Now the surgery obstruction $s(f)$ vanishes because it is a cobordism invariant. Let $i : M \subseteq W$ denote the inclusion. Then

$$
\begin{aligned}
\langle L(M) \cup f^* \sum_{j<i} \ell_j, [M] \rangle &= \langle i^* L(W) \cup i^* F^* \sum_{j<i} \ell_j, [M] \rangle \\
&= \langle L(W) \cup F^* \sum_{j<i} \ell_j, i_*[M] \rangle \\
&= 0 \;\; (\text{since } i_*[M] = 0) .
\end{aligned}
$$

Thus $\ell'$ vanishes on boundaries and is well-defined on $\Omega_{4i}(G/PL)$. Since $\ell'$ must map torsion to zero it induces a homomorphism

$$\underline{\ell}' : \underline{\Omega_{4i}(G/PL)} \longrightarrow \mathbb{Z}_{(2)} .$$

Theorem 4.5 states that $\underline{\ell}'$ induces a map

$$\ell : \underline{H_{4i}(G/PL)} \longrightarrow \mathbb{Z}_{(2)}$$

if and only if $\ell'$ vanishes on decomposables. We assume for the moment that $\ell'$ vanishes on decomposables. The universal coefficient theorem states that the evaluation map

$$H^{4i}(G/PL;\mathbb{Z}_{(2)}) \longrightarrow \mathrm{Hom}(H_{4i}(G/PL);\mathbb{Z}_{(2)}) \tag{2}$$

is onto. Thus there exists a cohomology class $\ell_i$ whose action on $H_{4i}(G/PL)$ is the composition

$$H_{4i}(G/PL) \longrightarrow \underline{H_{4i}(G/PL)} \xrightarrow{\ \ell\ } \mathbb{Z}_{(2)} .$$

The action of $\ell_i$ on $\Omega_{4i}(G/PL)$ is then exactly what is needed to satisfy (1) for manifolds of dimension $4k$, $k \leq i$.

**Proof that $\ell'$ vanishes on decomposables.** The decomposables of $\Omega_{4i}(G/PL;\mathbb{Z}_{(2)})$ are linear combinations of elements of the form

$$N_1^{4i-n} \times N_2^n \xrightarrow{\ p_2\ } N_2^n \xrightarrow{\ g\ } G/PL$$

where $n < 4i$ and $p_2$ is projection onto the second coordinate. To evaluate $\ell'$ on $[N_1 \times N_2, gp_2]$, we note that the $L$-genus is multiplicative and the Pontrjagin classes satisfy a Whitney sum formula modulo 2-torsion and so

$$L(N_1 \times N_2) = L(N_1) \times L(N_2)$$

modulo 2-torsion. Thus

$$\begin{aligned} \ell'[N_1 \times N_2, gp_2] &= \frac{s(gp_2)}{8} - \langle L(N_1) \times (L(N_2) \cup g^* \sum_{j<i} \ell_j), [N_1] \times [N_2] \rangle \\ &= \frac{s(gp_2)}{8} - \langle L(N_1), [N_1] \rangle \cdot \langle L(N_2) \cup g^* \sum_{j<i} \ell_j, [N_2] \rangle . \end{aligned} \tag{3}$$

First assume that $n \not\equiv 0 (\mathrm{mod}\, 4)$. Then $s(gp_2) = 0$ by the product formula for the index surgery obstruction of Rourke and Sullivan [11, Theorem 2.1]. Also, $\langle L(N_1), [N_1] \rangle = I(N_1) = 0$, so that both terms of (3) vanish and $\ell'[N_1 \times N_2, gp_2] = 0$.

Next assume $n \equiv 0 (\mathrm{mod}\, 4)$. If $n = 0$ then both terms of (3) are obviously zero. If $n > 0$ then

$$\frac{s(gp_2)}{8} = I(N_1) \cdot \frac{s(g)}{8}$$

by the product formula and

$$\langle L(N_1),[N_1]\rangle \cdot \langle L(N_2)\cup g^*\sum_{j<i}\ell_j,[N_2]\rangle \;=\; I(N_1)\cdot \frac{s(g)}{8}$$

by the inductive hypothesis ($N_2$ is a manifold of dimension $4j$ for some $j < i$). Thus $\ell'[N_1 \times N_2, gp_2] = 0$, and $\ell'$ vanishes on decomposables. The proof of Theorem 4.4 is complete.

**Remarks on Theorem 4.4.** (i) Since the evaluation map (2) has kernel a torsion group, the $\ell_i$ are unique up to the addition of torsion elements.
(ii) There are classes $L_i^{PL} \in H^{4i}(BPL;\mathbb{Q})$ which pull back to the $L$-genus in $BO$. (See Milnor and Stasheff [7].) The natural map $\pi : G/PL \longrightarrow BPL$ then satisfies

$$\text{(4)} \qquad \pi^*(L^{PL}-1) \;=\; 8\overline{\mathcal{L}} \qquad \text{(Sullivan [17], p.29)}$$

where $\overline{\mathcal{L}}$ denotes the image of $\mathcal{L}$ in $H^*(G/PL;\mathbb{Q})$.

**Proof of (4).** By our first remark we only need to verify that $\pi^*(L^{PL}-1)$ can be used to calculate the surgery obstruction for smooth manifolds. Let $M^{4k}$ be a $PL$ manifold, $f : M^{4k} \longrightarrow G/PL$ a map. The composition $\pi f$ is a stable $PL$ bundle over $M$. Then let $\nu_M$ be the stable normal bundle. We obtain a stable bundle $\nu_M - \pi f$ over $M$ and the fibre homotopy trivialization of $\pi f$ determines a normal invariant in $\pi_*(T(\nu_M - \pi f))$. The resulting surgery problem is the normal map associated to the map $f$. (See Chapter II.) The surgery obstruction of the map $f$ is thus equal to $[I(\nu_M - \pi f) - I(M)]$. The "index" of a stable bundle $\xi$ over $M^{4k}$ is defined by

$$I(\xi) \;=\; \langle L_k^{PL}(-\xi),[M^{4k}]\rangle \,.$$

Thus

$$\begin{aligned} s(f) &= \langle L_k^{PL}(\tau_M + \pi f),[M^{4k}]\rangle - I(M) \\ &= \langle L^{PL}(\tau_M)\cup L^{PL}(\pi f),[M^{4k}]\rangle - I(M) \\ &= \langle L^{PL}(M)\cup (L^{PL}(\pi f)-1),[M^{4k}]\rangle \end{aligned}$$

since $\langle L^{PL}(\tau_M),[M]\rangle = I(M)$. But $L^{PL}(\pi f) - 1 = f^*\pi^*(L^{PL}-1)$, so we have proved the desired formula for the surgery obstruction. We have also proved that Theorem 4.4 holds for $PL$ manifolds.
(iii) Let $M^{4k}$ be a manifold, smooth or $PL$. Then $[M, G/PL]$ forms a group via Whitney sum, and so it is natural to ask whether the surgery obstruction

$$s \;:\; [M,G/PL] \longrightarrow \mathbb{Z}$$

is a homomorphism. The answer is in general no. Since $L^{PL} \in H^*(BPL;\mathbb{Q})$ is multiplicative it follows from (4) that

$$h^*(\overline{\mathcal{L}}) \;=\; \overline{\mathcal{L}}\times 1 + 1\times\overline{\mathcal{L}} + \overline{\mathcal{L}}\times\overline{\mathcal{L}}$$

where $h : G/PL \times G/PL \longrightarrow G/PL$ is the multiplication in $G/PL$ induced by

Whitney sum. Thus if $fg \in [M, G/PL]$,

$$s(f \cdot g) \;=\; s(f) + s(g) + 8\langle L(M) \cup f^*\overline{\mathcal{L}} \cup g^*\overline{\mathcal{L}}, [M]\rangle \ .$$

(iv) By 3.8 the order of $x^{4i} \in H^{4i+1}(X_{4i-1}; Z)$ divides $\langle u, \alpha_i\rangle$ for any integral class $u \in H^{4i}(G/PL)$. Let $\nu_i$ denote the least positive integer such that $\nu_i \ell_i$ is integral. Then $\nu_i$ is of course always odd. We obtain a bound on $\nu_i$ as follows. By (ii) above $\pi^*(L^{PL} - 1) = 8\overline{\mathcal{L}}$. Let $\mu_i$ denote the least positive integer such that $\mu_i L_i^{PL}$ is integral. Brumfiel [2] has proved that

$$\mu_i \;=\; \prod_p \left[\frac{4i}{2(p-1)}\right]$$

where the product is taken over all odd primes $p \leq 2i+1$. Since $\nu_i$ divides $\mu_i$ we have: the order of $x^{4i}$ is a divisor of $\mu_i$. (The precise order of $x^{4i}$ can be computed using a result due to Dennis Sullivan, that $G/PL$ and $BO$ have the same homotopy type in the world of odd primes. It follows that the order of $x^{4i}$ is the odd part of $(2i-1)!$, for $i > 1$ .)

We conclude with a calculation of the fourth $k$-invariant $x^4 \in H^5(K(\mathbb{Z}_2, 2); \mathbb{Z})$ of $G/PL$. The following theorem is due to Sullivan [17].

**(4.6) Theorem.** $x^4 = \delta Sq^2 \imath_2$, *where* $\imath_2 \in H^2(K(\mathbb{Z}_2, 2); \mathbb{Z}_2)$ *is the fundamental class and* $\delta$ *is the Bockstein operation associated to the coefficient sequence*

$$\mathbb{Z} \xrightarrow{\times 2} \mathbb{Z} \longrightarrow \mathbb{Z}_2 \ .$$

**Proof.** Consider a section of the Postnikov system for $G/PL$

$$\begin{array}{ccc} G/PL & \longrightarrow & X_4 \\ & & \Big\downarrow K(\mathbb{Z},4) \\ & & X_2 = K(\mathbb{Z}_2, 2) \end{array}$$

By 3.8 the order of $x^4$ is the smallest positive integer $d$ such that there exists a cohomology class $u \in H^4(G/PL)$ satisfying $\langle u, \alpha_1\rangle = d$ , where $\alpha_1 \in \pi_4(G/PL)$ is a generator. By Theorem 4.4 there is a class $\ell_1 \in H^4(G/PL; \mathbb{Z}_{(2)})$ such that $\langle \ell_1, \alpha_1\rangle = 2$. Since there is an odd multiple of $\ell_1$ which is the reduction of an integral class, the order of $x^4$ divides an odd multiple of 2. But $x^4$ is in a 2-primary group. Thus $2x^4 = 0$. By the exactness of the sequence

$$H^4(K(\mathbb{Z}_2, 2); \mathbb{Z}_2) \xrightarrow{\delta} H^5(K(\mathbb{Z}_2, 2); \mathbb{Z}) \xrightarrow{\times 2} H^5(K(\mathbb{Z}_2, 2); \mathbb{Z})$$

there is a class $y \in H^4(K(\mathbb{Z}_2, 2); \mathbb{Z}_2)$ such that $\delta y = x^4$. But $H^4(K(\mathbb{Z}_2, 2); \mathbb{Z}_2) \cong \mathbb{Z}_2$ generated by $\imath_2^2 = Sq^2 \imath_2$ (Serre [13]), and so $x^4 = a\delta Sq^2 \imath_2$ for some $a \in \mathbb{Z}_2$. We complete the proof of Theorem 4.6 by showing that $x^4 \neq 0$.

The inclusion of the base point $i : * \subseteq G/PL$ induces a monomorphism

$$i_* \; : \; \Omega_4(*) \longrightarrow \Omega_4(G/PL)$$

and since the image of $i_*$ is in the kernel of the Hurewicz homomorphism, there is a diagram

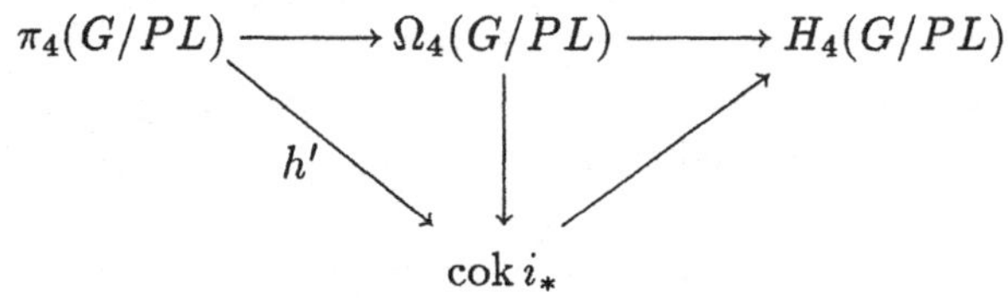


We have $\Omega_4(G/PL)/\text{torsion} = \mathbb{Z} \oplus \mathbb{Z}$, and so $\operatorname{cok} i_* \cong \mathbb{Z}\oplus$ finite group. Now the surgery obstruction $s : \Omega_4(G/PL) \longrightarrow \mathbb{Z}$ vanishes on $\operatorname{im} i_*$ and so induces a map $s' : \operatorname{cok} i_* \longrightarrow \mathbb{Z}$ such that

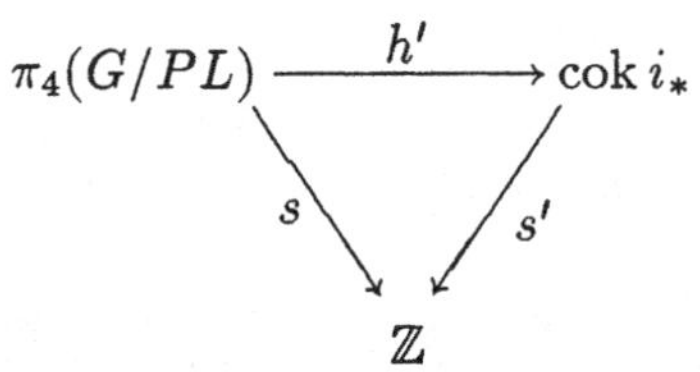


In order to prove that $x^4 \neq 0$, we show that $h' : \pi_4(G/PL) \longrightarrow \operatorname{cok} i_*$ is not an isomorphism onto a direct summand and apply 3.7. Since $\operatorname{cok} i_* \neq \mathbb{Z}\oplus$ finite group, we need only show that $\operatorname{im} s'$ properly contains $\operatorname{im} s$. Now $s(\alpha_1) = 16$, so that $\operatorname{im} s$ consists of multiples of 16. Thus it suffices to show

(*) there exists a map $f : \mathbb{CP}^2 \longrightarrow G/PL$ such that $s(f) = -8$.

**Proof of (*).** Let $\gamma$ denote the canonical complex line bundle over $\mathbb{CP}^2$. The total Chern class of $\gamma$ is $1 + x$, $x$ a generator of $H^2(\mathbb{CP}^2)$, and so the first Pontrjagin class $p^1(r\gamma)$ of the realification of $\gamma$ is $-x^2$. (The reader is referred to Milnor and Stasheff [7] for details.)

We show first that $24r\gamma$ is fibre homotopically trivial. The cofibration sequence

$$S^3 \xrightarrow{\ \eta\ } S^2 \longrightarrow \mathbb{CP}^2 \longrightarrow S^4 \xrightarrow{\ \Sigma\eta\ } S^3$$

induces an exact sequence

$$[S^3, BG] \xrightarrow{(\Sigma\eta)^*} [S^4, BG] \longrightarrow [\mathbb{CP}^2, BG] \longrightarrow [S^2, BG] \xrightarrow{\ \eta^*\ } [S^3, BG] \,.$$

We have

$$[S^3, BG] \cong \pi_2(G) \cong \pi_2^S = \mathbb{Z}_2 \text{ (generated by } \eta^2) \,,$$

$$[S^4, BG] \cong \pi_3(G) \cong \pi_3^S = \mathbb{Z}_{24} \text{ (generated by } \nu) \,.$$

Since $\eta^3 = 12\nu$ (for example, see Toda [21]) the cokernel of $(\Sigma\eta)^*$ is isomorphic to $\mathbb{Z}_{12}$. We also have

$$[S^2, BG] \cong \pi_1(G) \cong \pi_1^S = \mathbb{Z}_2 \text{ generated by } \eta \,,$$

so that $[S^2, BG]$ is generated by the Hopf bundle. Since the pullback of the Hopf bundle

$$\begin{array}{ccc} E & \longrightarrow & S^3 \\ {\scriptstyle \eta^*(\eta)}\downarrow & & \downarrow{\scriptstyle \eta} \\ S^3 & \xrightarrow{\eta} & S^2 \end{array}$$

is trivial, $\eta^* : [S^2, BG] \longrightarrow [S^3, BG]$ is the zero map and there is an exact sequence

$$0 \longrightarrow \mathbb{Z}_{12} \longrightarrow [\mathbb{CP}^2, BG] \longrightarrow \mathbb{Z}_2 \longrightarrow 0 \ .$$

Thus $[\mathbb{CP}^2, BG]$ is a group of 24 elements and $24r\gamma$ is fibre homotopically trivial.

The composite

$$\mathbb{CP}^2 \xrightarrow{24r\gamma} BO \longrightarrow BPL \longrightarrow BG$$

is trivial and so the associated $PL$ bundle

$$\xi \ : \ \mathbb{CP}^2 \longrightarrow BPL$$

factors through $G/PL$:

$$\begin{array}{ccc} \mathbb{CP}^2 & \xrightarrow{24r\gamma} & BO \\ {\scriptstyle f}\downarrow & & \downarrow \\ G/PL & \xrightarrow{\pi} & BPL \end{array}$$

We calculate $s(f)$ using the remarks following Theorem 4.4

$$\begin{aligned} s(f) &= \langle L(\mathbb{CP}^2) \cup (L^{PL}(\pi f) - 1), [\mathbb{CP}^2] \rangle \\ &= \langle (1 + L_1(\mathbb{CP}^2)) \cup L_1^{PL}(\pi f), [\mathbb{CP}^2] \rangle \\ &= \langle L_1^{PL}(\pi f), [\mathbb{CP}^2] \rangle \\ &= \langle \frac{p_1(24r\gamma)}{3}, [\mathbb{CP}^2] \rangle \\ &= \langle \frac{-24x^2}{3}, [\mathbb{CP}^2] \rangle \\ &= -8 \ . \end{aligned}$$

This completes the proof of $(*)$ and Theorem 4.6 follows.

## References

[1] W. Barcus and J.-P. Meyer, **On the suspension of a loop space**, Amer. Jour. Math. 80 (1958) 895–920.

[2] G. Brumfiel, **On integral $PL$ characteristic classes and realizing homology classes by manifolds**, Topology 8 (1969) 39–46.

[3] P. E. Conner and E. E. Floyd, **Differentiable periodic maps**, Ergebnisse der Mathematik 33, Springer (1964).

[4] S. Eilenberg and S. MacLane, **Relations between homology and homotopy groups of spaces II**, Annals of Math. 51 (1950) 514–533.

[5] F. Hirzebruch, **Topological methods in algebraic geometry**, Grundlehren der mathematischen Wissenschaften 131, Springer (1966).

[6] D. W. Kahn, **Induced maps for Postnikov systems**, Trans. Amer. Math. Soc. 107 (1963) 432–450.

[7] J. Milnor and J. D. Stasheff, **Characteristic classes**, Annals of Math. Studies 76, Princeton University Press, Princeton (1974).

[8] J. C. Moore, **Systémes de Postnikov et complexes monoidaux**, Exposé 21, Seminaire Cartan, 1954–55.

[9] J. C. Moore, **Semi-simplicial complexes and Postnikov systems**, Symposium Internacional de Topologia Algebraica, Universidad Nacional de Mexico and UNESCO, Mexico City (1958).

[10] M. M. Postnikov, **Investigations in the homotopy theory of continuous mappings, I, II, III**, Translations Amer. Math. Soc. 7 (1957) 1–134 and 11 (1959) 115–153.

[11] C. P. Rourke and D. P. Sullivan, **On the Kervaire obstruction**, Annals of Math. 94 (1971) 397–413.

[12] J.-P. Serre, **Homology singulière des espaces fibrés**, Annals of Math. 54 (1951) 425–505.

[13] J.-P. Serre, **Cohomologie modulo 2 des complexes d'Eilenberg-Mac Lane**, Comm. Math. Helv. 27 (1953) 198–232.

[14] E. H. Spanier, **Algebraic topology**, McGraw-Hill, New York (1966).

[15] R. E. Stong, **Notes on cobordism theory**, Mathematical Notes 7, Princeton University Press, Princeton (1968).

[16] D. P. Sullivan, **Triangulating homotopy equivalences**, Thesis, Princeton University (1966).

[17] D. P. Sullivan, **Triangulating and smoothing homotopy equivalences and homeomorphisms, Geometric Topology Seminar Notes**, Princeton (1967). (Published in this volume).

[18] D. P. Sullivan, **On the Hauptvermutung for manifolds**, Bull. Amer. Math. Soc. 73 (1967) 598–600.

[19] H. Suzuki, **On the Eilenberg-MacLane invariants of loop spaces**, Jour. Math. Soc. Japan 8 (1956) 93–101.

[20] E. Thomas, **Seminar on fibre spaces**, Lecture Notes in Mathematics 13, Springer (1966).

[21] H. Toda, **Composition methods in homotopy groups of spheres**, Annals of Math. Studies 49, Princeton University Press, Princeton (1962).

[22] G. W. Whitehead, **On the homology suspension**, Annals of Math. 62 (1955) 254–268.

Institute for Advanced Study
Princeton
Spring 1968

# Coda: connection with the results of Kirby and Siebenmann

**By C. P. Rourke**

We have shown that $TOP/PL \longrightarrow G/PL$ factors via

$$K(\mathbb{Z}_2, 3) = \text{fibre}(G/PL \longrightarrow \Omega^{4n}(G/PL)) .$$

Now Kirby and Siebenmann have shown [2, 3, 4] that $TOP/PL$ is also a $K(\mathbb{Z}_2, 3)$, and that the map $q_h : M \longrightarrow TOP/PL$ is the obstruction to an isotopy from $h$ to a $PL$ homeomorphism. There are two possibilities:

(1) $TOP/PL \longrightarrow K(\mathbb{Z}_2, 3)$ is null-homotopic;

(2) $TOP/PL \longrightarrow K(\mathbb{Z}_2, 3)$ is a homotopy equivalence.

We shall eliminate (1). Both the fibration and the theorem mentioned in the introduction then follow if we combine the Kirby-Siebenmann result with the main theorem of Chapter III. In order to eliminate (1) it is necessary to consider the structure sequence for the torus $T^r$. There is a fibration onto its image

$$HT(T^r) \longrightarrow (G/PL)^{T^r} \longrightarrow \mathbb{L}_{r+4n}(\mathbb{Z}^r)$$

due to Casson and Quinn [5]. Now $\mathbb{L}_{r+4n}(\mathbb{Z}^r)$ consists of $(\Delta^k, r+4n)$-oriented normal maps (which are homotopy equivalences on boundaries) together with a reference map to a $K(\mathbb{Z}^r, 1)$, which we can take to be $T^r$ itself. Consequently there is a map

$$\alpha : \mathbb{L}_{4n}(\{1\})^{T^r} \longrightarrow \mathbb{L}_{r+4n}(\mathbb{Z}^r)$$

defined as follows. Let $f : T^r \longrightarrow \mathbb{L}_{4n}(\{1\})$ be given; then $f$ determines an $(i+4n)$-normal map for each $i$-simplex of $T^r$ and, glueing together, we obtain an $(r+4n)$-normal map over $T^r$, in other words a simplex of $\mathbb{L}_{r+4n}(\mathbb{Z}^r)$. Using the Splitting Theorem of Farrell [1] we can convert any normal map (homotopy equivalence on boundary) over $T^r$ into an assemblage of normal maps (homotopy equivalences on boundaries) over simplexes of $T^r$. This argument generalizes to show that $\alpha$ is a homotopy equivalence. Now $\mathbb{L}_{4n}(\{1\})$ and $\Omega^{4n}(G/PL)$ have the same homotopy type, by considering the structure sequence for $D^{4n}$ rel $\partial$, and we can rewrite our fibration as

$$HT(T^r) \longrightarrow (G/PL)^{T^r} \longrightarrow (\Omega^{4n}(G/PL))^{T^r} .$$

It follows that $HT(T^r)$ and $(K(\mathbb{Z}_2, 3))^{T^r}$ have the same homotopy type. Now if the map $TOP/PL \longrightarrow K(\mathbb{Z}_2, 3)$ is null-homotopic, then any self-homeomorphism of $T^r$ is homotopic to a $PL$ homeomorphism. However Siebenmann [2, 3, 4] has constructed a self-homeomorphism of $T^6$ which is **not** homotopic to a $PL$ homeomorphism. Hence

$$TOP/PL \longrightarrow K(\mathbb{Z}_2, 3)$$

must be a homotopy equivalence.

It is clear from the above discussion that any homotopy equivalence onto $T^r$

is homotopic to a homeomorphism, and that the obstructions to the homotopy and isotopy Hauptvermutung coincide for $Q = T^r$. This contrasts with the simply connected case and shows that the general solution is bound to be somewhat complicated.

## References

[1] F. T. Farrell, **The obstruction to fibering a manifold over a circle**, Ph.D. Thesis, Yale (1967) and Bull. Amer. Math. Soc. 73 (1967) 737–740.

[2] R. C. Kirby, **Lectures on the triangulation of manifolds**, mimeographed notes, U.C.L.A. (1969).

[3] R. C. Kirby and L. C. Siebenmann, **On the triangulation of manifolds and the Hauptvermutung**, Bull. Amer. Math. Soc. 75 (1969) 742–749.

[4] R. C. Kirby and L. C. Siebenmann, **Foundational essays on topological manifolds, smoothings, and triangulations**, Annals of Math. Studies 88, Princeton University Press (1977).

[5] F. S. Quinn, **A geometric formulation of surgery**, Ph.D. Thesis, Princeton (1969) and pp. 500–511 of Topology of Manifolds, Ed. J. C. Cantrell and C. H. Edwards, Markham, Chicago (1970).

Warwick University
1972

## *K*-Monographs in Mathematics

1. A.A. Ranicki (ed.), A.J. Casson, D.P. Sullivan, M.A. Armstrong, C.P. Rourke and G.E. Cooke: *The Hauptvermutung Book.* A Collection of Papers on the Topology of Manifolds. 1996 ISBN 0-7923-4174-0
2. L. Huichi and F. van Oystaeyen: *Zariskian Filtrations.* (*not yet published*)

KLUWER ACADEMIC PUBLISHERS – DORDRECHT / BOSTON / LONDON

The manufacturer's authorised representative in the EU is Springer Nature Customer Service Centre GmbH, Europaplatz 3, 69115 Heidelberg, Germany. If you have any concerns regarding our products, please contact ProductSafety@springernature.com

Printed and bound by CPI Group (UK) Ltd, Croydon, CR0 4YY

07/01/2026

02030929-0001